Lake Urmia's Hydro-Social Dilemma

Robert Gonda

Lake Urmia's Hydro-Social Dilemma

From Turquoise Solitaire to White Desert

palgrave
macmillan

Robert Gonda
Augsburg, Germany

ISBN 978-3-032-08940-3 ISBN 978-3-032-08941-0 (eBook)
https://doi.org/10.1007/978-3-032-08941-0

Simultaneously dissertation, University of Augsburg, 2025. D384

This Palgrave Macmillan imprint is published by the registered company Springer Nature Switzerland AG
The registered company address is: Gewerbestrasse 11, 6330 Cham, Switzerland

Acknowledgments I would like to thank **Prof. Dr. Matthias Schmidt**, Chair of Human Geography and Transformation Research at the University of Augsburg, for his supervision and constructive feedback during the course of this research. His trust and patience in my abilities have been invaluable.

My thanks also go to **Dr. Nosrat Heidari** and to the **University of Urmia (Iran)**, whose assistance and collaboration on the ground opened doors that would otherwise have remained closed. I am grateful to my translator, whose skill and dedication made fieldwork not only possible but truly meaningful.

Special thanks are due to the student assistants whose work and commitment greatly contributed to this research. I am also grateful to my colleague **Sebastian Transiskus**, who joined me during fieldwork for his own research and with whom I enjoyed many valuable debates about Iran and our respective projects. I will treasure our time in Iran as a lasting and precious memory.

This research was made possible through the financial support of the **Deutsche Forschungsgemeinschaft (DFG, German Research Foundation)**, which funded the project from 2020 to 2022.

I am indebted to all interview participants at Lake Urmia for their warmth, openness, and remarkable hospitality, often extended under challenging circumstances. Their voices, stories, and trust are at the heart of this work.

Finally, I thank my friends and family for their patience, encouragement, and quiet faith throughout the long journey of this dissertation. I am deeply grateful to my mother for her unwavering support, to my brother for his help and advice, and to my partner, who has often listened to my theoretical musings—especially my recurring question, posed at least once a month over many years: What is water? Without her patience and love, I could not have seen this project through.

Competing Interests The author has no competing interests to declare that are relevant to the content of this manuscript.

Ethics Approval This study was conducted in accordance with the principles of the 1964 Declaration of Helsinki and its subsequent amendments. No formal institutional ethics review process was applicable for this research context. Informed consent to participate was obtained orally from all participants prior to the interviews. Written consent was not sought, as the research took place in a politically sensitive environment in the Islamic Republic of Iran, where signed documents containing identifying information could have posed significant risks to both participants and the research team. The interviews were conducted using a standardised questionnaire, administered by a local interpreter who translated both questions and answers. To safeguard anonymity and minimise potential harm, no personal identifiers were collected, no audio or video recordings were made, and responses were documented solely through contemporaneous note-taking by the researcher. All participants were informed about the aims of the study, the voluntary nature of their participation, and their right to withdraw at any point without consequence.

Contents

About the Author

Robert Gonda is a researcher in human geography and political ecology. He earned his Ph.D. from the University of Augsburg, where he investigated the ecological crisis of Lake Urmia in Iran through hydro-social and Latourian perspectives. His research explores the intersections of science, society and the environment, with a particular focus on water governance and socio-ecological controversies.

List of Figures

List of Tables

CHAPTER 1

White Is the New Turquoise

The turquoise solitaire of north-west Iran, Lake Urmia, has mutated. Where once ten thousand flamingos flew to breed, tourists swam during vacations and poems were written about the lake's aesthetics, there is now nothing but a mere memory of those times. Salt plains, salt winds, environmental degradation, droughts, water scarcity, rural poverty, rural exodus, loss of livelihoods and decreasing biodiversity are the new descriptions and terms associated with Lake Urmia, next to many others. The three provinces of West Azerbaijan, East Azerbaijan and Kurdistan, which are located in the 52,000 km^2 basin, face increasing challenges accompanying this socio-environmental disaster. Over roughly two decades, the lake that had once covered an area of more than 5,000 km^2 in the early to mid-1990s has now almost vanished (Figs. 1.1 and 1.2) or, as mentioned above, mutated. Admittedly, 'mutation' is not a typical description for a desiccating waterbody, a hypersaline lake to be precise, reminiscent of the Aral Sea and its pertinent desert, the Aralkum. Nevertheless, it is the word I have chosen to move my investigation into the waters of Bruno Latour's body of work and a theory—or paradigm—called the 'hydro-social cycle'.

Lake Urmia proves that it is imperative to question the modern understanding of water and the concept of nature itself. No longer does it suffice to blame natural or anthropogenic drivers for such a disaster, since the line between what is nature and what is anthropogenic has

R. Gonda, *Lake Urmia's Hydro-Social Dilemma*,
https://doi.org/10.1007/978-3-032-08941-0_1

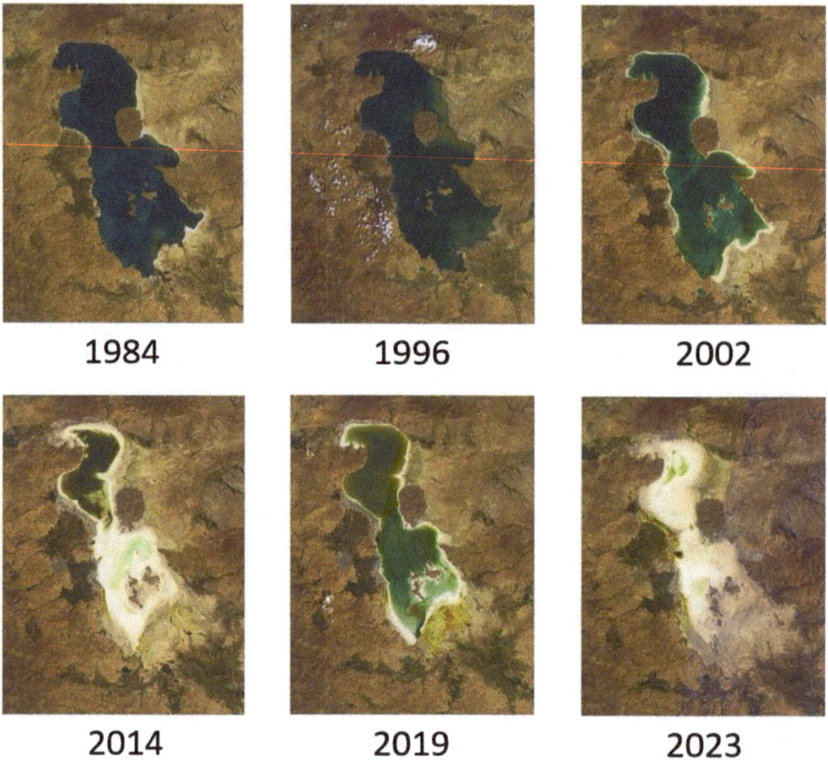

Fig. 1.1 Lake Urmia's development from 1984 to 2023 (September months compared; Landsat)

blurred significantly in the face of *anthropogenic* climate change. Furthermore, it is no longer sufficient to tell coherent after-the-fact stories in times of *controversies* that short-circuit a complex set of reasons for how such a phenomenon could occur, and it is no longer enough to assess the current state of an environment by imposing concepts on it that are presented as *facts*, even though they are anything but ahistorical or irrefutably objective. What is needed, in fact, is a new concept that breaks with the traditional human–nature dichotomy—an investigation that tends to dwell in the 'messiness' of (sometimes) divergent viewpoints, i.e., a historicisation of facts and an emphasis on processes. Mutation may be just a fancy word for 'change'; however, it marks an important point,

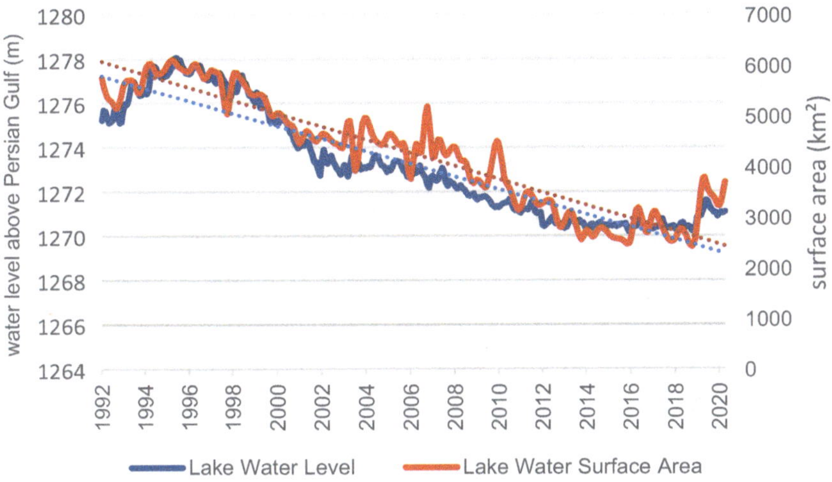

Fig. 1.2 Water level and surface area of Lake Urmia from 1992 to 2021 (own illustration based on Sima et al. [2021]; database from dimensions.ai)

in that it is not only Lake Urmia that has been affected by change, but also numerous entities.

However, what has led to Lake Urmia's mutation? This is a vexing question, and one which I am not willing to answer, at least for now. Conversely, I am more interested in how this question and the solutions as well as different concepts have been developed over time to provide answers to this phenomenon. As investigating how Lake Urmia's demise has been explained, it allows us to explore differences between diverging points of view—the same goes for the solutions provided to restore the lake. This is the main point of this dissertation: How do (or have) different actors perceived the metamorphosis of Lake Urmia? What is known, nevertheless, is that the alteration of the lake has been and is going to be profound, as if the 'new' Lake Urmia has almost nothing in common with the 'old' version. The mutation that I have declared with such certainty is one of the preliminary conclusions of my thesis. However, the recent history of Lake Urmia and the analysis of its discourse reveal numerous opinions on the matter. It is my intention to elucidate this spectrum of views and to highlight controversial issues. The focus is on the scientific discourse, while the opinions of other actors, such as farmers, ministries or other experts, have been included as supplementary evidence. As such, this thesis can be categorised as a review of the (scientific) discourse on Lake Urmia.

1.1 Research Questions and Thesis Organisation

As mentioned in the introduction, this thesis's overarching goal is to illuminate and reflect on the social, primarily scientific discourse on Lake Urmia in the last ca twenty years. The modus operandi will be two-fold. On the one hand, I have chosen the hydro-social cycle, Latour's works on the human–environment relationship and water ethics, and on the other hand the cartography of controversy, as tools to investigate the discourse systematically. It is against this backdrop that I ultimately wish to answer the following:

- What is driving hydro-social dynamics at Lake Urmia?

Since I have not explained the hydro-social cycle yet, which might be an issue for those unfamiliar with the concept, I rephrase the question in more general terms: 'What is driving the human-water relationship at Lake Urmia?' However, there is a significant difference between these two questions, as will be made clear in the conclusion to this dissertation. For now, though, it fulfils its purpose.

Furthermore, to answer this central research question from a social scientist's angle, I have chosen to approach this investigation by asking the following question:

- What actors are held accountable for the Lake Urmia crisis according to the social/scientific discourse, and which ones take a central position?

Obviously, there exists a long list of responsible actors; however, I am primarily interested in those controversially discussed. As such:

- What topics are heavily debated, and which concepts or ideas on the nature of this crisis could these conflicts reveal about hydro-social dynamics at Lake Urmia?

Lake Urmia, however, is not an isolated phenomenon in Iran (Madani, 2014; Madani et al., 2016), as its history is also embedded in a broad historical context of the country's water management policy, which has to be explored as well. This in turn helps gain a sense of how the human-water relationship has changed over time and what sort of dynamics are at work in the Lake Urmia basin. Therefore, the following question has been formulated:

- How have hydro-social concepts (human-water relationship) in Iran transformed from their inception to the present, and what effects could they have on Lake Urmia?

Seven chapters help answer these overarching questions. In Chapter 2, I set the tone for this thesis by introducing the major themes relevant to the hydro-social cycle and some of Latour's ideas on the role of sciences in the production of knowledge and his perspectives on the—in his view, obsolete—society–nature dichotomy. The focal point here is set on different understandings of water and water crises and to question the 'nature', or ontology, of water per se. The chapter ends by suggesting a pathway to investigate water crises through the lens of controversy studies, as proposed by Venturini and Munk (2022).

Thereafter, I review and analyse the history of Iran's water management. Four different periods were chosen for the investigation, and these demonstrate how water has been conceptualised and treated differently over time. Moreover, the politics of water, with all its ideologies and particularities, show how Lake Urmia might be just a symptom at the tail end of long and ongoing processes.

The historical chapter is followed by a systematic presentation of the scientific landscape of Lake Urmia. The agenda of Chapter 4 is to determine the feasibility of Lake Urmia as a controversial study, to historicise scientific history in congruence with the decline of the lake, to illustrate how the scientific community has approached the matter, what issues it has focused on and which actors might be of major importance in the production of knowledge. Finally, it identifies the main controversies that are of interest to this study.

Chapters 5, 6, and 7 focus on these controversies, namely the effects of a causeway constructed across the lake, diverging opinions on the minimal ecological water level that was defined by a group of scientists, i.e., the main cause of the disaster (human vs. natural), and last but not least the different solutions proposed by different groups to solve the crisis. However, it should be noted that other than 'just' describing these controversies, I approach these topics from different—sometimes critical—angles to scrutinise entrenched 'truths' or objective facts.

The thesis ends with a conclusion followed by a discussion on both theory, where I disclose my thoughts on the hydro-social cycle, and the empirical case for Lake Urmia.

1.2 State of the Art at Lake Urmia

In general, there is a significant quantitative discrepancy between natural and social sciences when conducting research on Lake Urmia. The natural sciences are hegemonic in this case, and only a small number of papers have been written on the matter by social scientists—the dangers of this inconsistency are analysed and discussed in a later chapter of this dissertation. Moreover, in this sense, the contribution of social sciences is highlighted and summarised in order to situate this thesis.

A common denominator of the following summary of academic papers is their emphasis on the influence of politics—regional, national or geopolitical—on Lake Urmia's development. Furthermore, they highlight the social dynamics that have been mostly obfuscated by the majority of the literature. However, this perspective is rarely chosen, so I have compressed the major findings of these social studies.

One of the first extensive PhDs to present the multifaceted dimensions of Lake Urmia's demise was undertaken by Hashemi (2012), who comprehensively assessed the socio-technological from with an in-depth-perspective, since he was indeed involved in a variety of decision-making processes. In detail, he disclosed organisational water management structures, reported on numerous technical and political decision-making processes and offered an integrated water resource management plan as a sustainable solution to the Lake Urmia crisis. Other papers focused on the political context, given both in geopolitical and regional terms. While Khalyani et al. (2014) accentuated power imbalances within the basin between different ethnicities (Kurds and Azeri), Dalby and Moussavi (2016) related Lake Urmia to the geopolitical dynamics of Iran and its internal/external challenges which in part led to the crisis. In her master's thesis, Rouw (2017) analysed the social debate around the lake by identifying different discursive frameworks, such as the agricultural, political and natural frames, albeit she did not include the scientific discourse. Similarly, Ranjbar (2022) unearthed a hidden narrative relating to local protesters to provide insights into the fragile relationship between the basin's population and the central government.

During the last decade, recent scholarship has shed light on the multiple motivations of human actors in order to explain adequately the primary causes of this ongoing crisis. Farmers, institutions such as the ULRP (Urmia Lake Restoration Program), the Ministry of Energy and the Ministry of Agriculture are increasingly in the spotlight, and discussions have shifted from simple objects and things made to act (actants, e.g., dams and crops), to concrete human actors who make other entities act. In other words, socio-political investigations have gained momentum in recent years. For instance, Salimi et al. (2019) criticised the ULRNC (Urmia Lake Restoration National Committee) and its members because any efforts to restore the lake were hindered by contradictory national policy agendas and agricultural development.

Esmailzadeh et al. (2023), who analysed the issue from the perspective of the three relevant ministries (Department of Environment, Ministry of Agriculture and Ministry of Energy), formulated a similar critique, concluding that there was a lack of awareness and insight into the problems on the ground, i.e., the socio-economic struggles of the farmers in particular. In addition to institutions, the motives of farmers, who consumed most of the water in the basin, remained largely unknown. This uncertainty, however, has been addressed by authors such as Pouladi et al., (2020, 2021), Ženko in her cumulative dissertation (2020), Farahmand et al. (2023) and Abbaszadeh et al. (2020). Their findings suggest that farmers prioritise economic, vested interests in their land over social norms such as water conservation and conceptualise the lake's water cycle differently, thus downplaying the disaster per se, which overall contradicts, for example, the ULRP's goal to successfully restore the lake.

Moreover, the perspective on (political) power highlights asymmetries between local Kurd and Azeri ethnic groups, as exemplified by access to and the availability of water (worse for the former and better for the latter); the desiccation of Lake Urmia causing mental health issues for local dwellers, a phenomenon described as 'solastalgia'; the changing narrative in water governance; different meanings and values attributed to water for downstream users; Finally, there are deficits in integrated water resource management, with a lack of social impact assessment

(Ženko, 2020; Ženko & Ulezic, 2019). The aforementioned field studies conducted more than 800 in-depth interviews, allowing researchers—and perhaps even officials—to register meaningful insights for further exploration of the socio-political nature of the Lake Urmia disaster. In addition, Ženko (2020) proposes, among other things, to study discursive power strategies and local water governance alongside traditional ones.

To date, only two other studies have attempted to synthesise the scientific findings of the past 20 years (Garousi et al., 2013; Parsinejad et al., 2022). The review article by Garousi et al. (2013) analysed 36 selected articles to summarise the possible causes of the decline of Lake Urmia, alongside negative impacts, the consequences thereof and possible solutions. However, since hundreds of new articles have been published since 2013, the findings of Garousi et al. (2013) are limited to the small number of papers they were able to include at the time. Moreover, the analysis of the selected literature remained superficial, numerical and, most importantly, uncritical.

The study by Parsinejad et al. (2022) is the latest (and only second) review article, albeit in much larger quantities. In contrast to Garousi et al. (2013), it takes a natural science perspective, evaluating the methods and results of several hundred peer-reviewed articles. My work differs in the sense that it focuses on qualitative differences and controversies and offers a critical evaluation of naturalised concepts, following, among others, the hydro-social cycle framework.

In the context of Lake Urmia, only Ženko (2020) and Abbaszadeh et al. (2020) have applied the hydro-social cycle as a paradigm; however, the latter was published only in the Persian language, except for the abstract. Ženko (2020) focused on the unequal vulnerability of ethnic minorities in the region, the relation between water scarcity and mental health issues and the anthropogenic influence that resulted in Lake Urmia becoming what it is today. Abbaszadeh et al. (2020) published a paper in Farsi, highlighting farmers' perspectives on the water crisis in each of the three aforementioned provinces.

Latour's work as a thought-provoking impulse, however, has not been applied by any study dealing with Lake Urmia. Similarly, a controversy focused study inspired by the Actor–Network Theory and controversy mapping (Venturini & Munk, 2022) has not been attempted by any researcher, either.

1.3 Methodology

A total of 82 in-depth interviews were conducted, 77 of which were held mainly with farmers from 41 villages in the vicinity of Lake Urmia—within a maximum distance of 10 km. Of these villages, 89% (68) were located in West Azerbaijan and 11% (9) in East Azerbaijan. The interviews were conducted in two phases. The majority (69) took place during my field visit from 9 to 30th September 2019, with the help of a local interpreter fluent in English and several local languages (Kurdish, Farsi and Azeri) and a driver who enabled us to reach even remote villages. Sites were selected primarily based on the experience of an initial, exploratory field visit from September 29th to 7th October 2018. The interviews were bilateral, but in some cases the interviewees were accompanied either by their families or by other farmers, whose insights were recorded if they were considered useful to the research. Participants were randomly selected; the only requirement was that they were farmers or at least had some connection to the agricultural sector (e.g., local water manager, village councillor, village elder). Otherwise, my interpreter and I approached our interviewees without any ulterior motives. The purpose of my research was explained to the participants, and we assured them that I was not an official of the Iranian government; some of them wanted to see documents to prove the purely scientific nature of my research. With this in mind, I guaranteed the anonymity of all my interviewees.

The second set of interviews (eight) with local farmers was conducted by my local supervisor from June to July 2022, because during the Covid-19 pandemic, Iran put in place severe restrictions in relation to both entry into the country and social interaction. For this reason, and in combination with the politically tense situation, I did not undertake another field trip.

Similarly, four of my five expert interviews, conducted from June to July 2021, were carried out by email and one was a live interview. Again, the political situation in Iran, especially since 2020, and the political nature of my research made these experts reluctant to engage in a live conversation with me via Skype or Zoom, instead preferring to answer my questions in writing.

The primary aim of the interviews was to gain an understanding of the Lake Urmia crisis from different—and especially qualitative—viewpoints. Therefore, the questions generally focused on the causes of the disaster, perceptions, possible solutions and consequences, especially for farmers.

Thirteen (16%) of my interview partners were female and the other 69 (84%) male. I am well aware of the gender bias that may result from such a one-sided pool of participants. Although, occasionally, women in a leading position within the village community could be interviewed, men usually took the foremost role in the interviews in the family nucleus. This cultural constraint has also been recognised by Ženko (2020:15–16).

Other than the interviews, and as mentioned above, I focused on a literature review, which consisted of 115 documents, including five official reports produced by relevant organisations and institutions. The goal here was to highlight the differences between scientific viewpoints and to establish a heterogeneous landscape of the academic landscape. As such, the scientific articles were selected at first systematically—as explained in detail in Chapter 4—and, as a profound text corpus was established, other articles were added to further deepen the investigation of the controversies. Search engines such as dimensions.ai and the freeware software Harzing's publish or perish were used to find articles based on several keywords, including 'Lake Urmia' or 'Lake Orumiyeh' and specific keywords for the designated controversy.

The interviews were transcribed and together with the primarily (scientific) literature coded and analysed with MAXQDA. The codes were based on theory (deductive) and the text corpus (inductive) (Glasze & Mattissek, 2021). Although I chose an interpretative—and thus less rigid—discourse analysis, I do disclose my reasoning why I coded certain passages the way I did. In this sense, I wish to prevent any notion that my interpretation may appear arbitrary or that the reader is being forced to trust my insights without any empirical evidence.

References

Abbaszadeh, M., Banifatemeh, H., & Nikdel, N. (2020). A Study of farmer's perception of Urmia Lake (UL) crisis with grounded theory approach. *Journal of Applied Sociology, 31*(2), 1–4.

Dalby, S., & Moussavi, Z. (2016). Environmental security, geopolitics and the case of Lake Urmia's disappearance. *Global Change, Peace & Security, 29*(1), 39–55.

Esmailzadeh, S., Rizi, A. P., & Mianabadi, H. (2023). Evaluation of the water policies of the Urmia Lake Basin: Has the government accurately identified the problem? *Environment: Science and Policy for Sustainable Development, 65*(6), 18–34.

Farahmand, H., Tajrishy, M., Isaai, M. T., Ghoreishi, M. & Mohammadi, M. (2023). *How do farmers' perceptions and attitudes toward agricultural water consumption behaviors can lead to unsustainability; evidence from Mahabad plain*, Lake Urmia, Iran. https://www.researchsquare.com/article/rs-2478328/v1. Accessed September 21, 2024.

Garousi, V., Najafi, A., Samadi, A., Rasouli, K., & Khanaliloo, B. (2013). *Environmental crisis in Lake Urmia, Iran: A systematic review of causes, negative consequences and possible solutions*. Unpublished.

Glasze, G., & Mattissek, A. (2021). *Handbuch Diskurs und Raum*. transcript Verlag, Bielefeld, Germany.

Hashemi, M. (2012). *A social-technical assessment framework for integrated water resources management (IRWM) in Lake Urmia Basin, Iran*. Doctor of Philosophy Thesis. Newcastle University, Newcastle.

Khalyani, A. H., Mayer, A. L., & Norman, E. S. (2014). Water Flows Toward Power: Socioecological Degradation of Lake Urmia, Iran. *Society & Natural Resources, 27*(7), 759–767.

Madani, K. (2014). Water management in Iran: What is causing the looming crisis? *Journal of Environmental Studies and Sciences, 4*(4), 315–328.

Madani, K., AghaKouchak, A., & Mirchi, A. (2016). Iran's Socio-economic drought: Challenges of a water-bankrupt nation. *Iranian Studies, 49*(6), 997–1016.

Parsinejad, M., Rosenberg, D. E. & Ghale, Y. A. G. et al. (2022) 40-years of Lake Urmia restoration research: Review, synthesis and next steps. The Science of the total environment (p. 832).

Pouladi, P., Afshar, A., Molajou, A., & Afshar, M. H. (2020). Socio-hydrological framework for investigating farmers' activities affecting the shrinkage of Urmia Lake; hybrid data mining and agent-based modelling. *Hydrological Sciences Journal, 65*(8), 1249–1261.

Pouladi, P., Badiezadeh, S., & Pouladi, M. et al. (2021). Interconnected governance and social barriers impeding the restoration process of Lake Urmia. *Journal of Hydrology, 598*.

Ranjbar, A. M. (2022). The greening of human rights in Iran: Lake Orumiyeh, human rights, and environmental justice. *Environment and Planning e: Nature and Space, 3*(5), 251484862211081.

Rouw M. (2017). Socially navigating through the Urmia Lake debate: Master Thesis. https://edepot.wur.nl/421717. Accessed August 15, 2024.

Salimi, J., Maknoon, R., & Meijerink, S. (2019). Designing institutions for watershed management: A case study of the Urmia Lake Restoration National Committee. *Water Alternatives, 12*(2), 609–635.

Sima, S., Rosenberg, D. E., Wurtsbaugh, W. A., Null, S. E. & Kettenring, K. M. (2021). Managing Lake Urmia, Iran for diverse restoration objectives:

Moving beyond a uniform target lake level. *Journal of Hydrology: Regional Studies 35*(9), 100812.

Venturini, T., & Munk, A. K. (2022).*Controversy mapping: A field guide*. Polity Press, Cam-bridge, UK, Medford, MA, USA.

Ženko, M. (2020). *Hydro-social transformations in the Lake Urmia basin, Iran*. https://ddd.uab.cat/pub/tesis/2021/hdl_10803_670989/maze1de1.pdf. Accessed September 21, 2024.

Ženko, M., & Ulezic, S. (2019). The unequal vulnerability of Kurdish and Azeri minorities in the case of the degradation of Lake Urmia, Iran. *Journal of Political Ecology, 26*, 167–183.

CHAPTER 2

The Social Power of Water: How the Hydrologic Cycle Turns Against Modern(ising) Societies

The water-human relationship is as diverse as it is paradoxical. It is not far-fetched to assume that most people regard water as the source of life. Climate change, however, is increasingly turning this precious element into either a dangerous threat or a scarce resource worthy of war—at least according to some actors. Hydrologic events such as droughts, floods, tsunamis, cyclones, storm surges and convective storms account for 90% of all human disasters (Seung-soo, 2018). The negative socio-economic impact of water-related disasters is therefore not short of depressing figures. From 1995 to 2015, for instance, droughts and floods killed 166,000 people and destroyed property with an estimated value of around US$700 billion (UN, 2020: 24). Droughts in particular have become a major concern for human development in many parts of the world. According to Mekonnen and Hoekstra (2016: 3), four billion people experience conditions of severe water scarcity for at least one month per year—a situation the authors can only describe as 'worse than suggested by previous studies'. Apparently, the water crisis is a global phenomenon, exacerbated by anthropogenic climate change, and consequently water frequently oscillates from destructive abundance to intensifying scarcity. The middle ground, where water could be called rightfully 'the source of life', is shrinking.

So far, water's power over humanity has been emphasised, and therefore the impression is that water—in a deterministic fashion—tightly

R. Gonda, *Lake Urmia's Hydro-Social Dilemma*,
https://doi.org/10.1007/978-3-032-08941-0_2

grips human civilisations. In stark contrast, the Anthropocene proposes a converse relationship, whereby humans increasingly manipulate and control the hydrologic cycle to such a degree that, in the words of Steffen et al. (2011: 843), '... humankind, our own species, has become so large and active that it now rivals some of the great forces of Nature in its impact on the functioning of the Earth system'. According to the Earth System Science, this human impact calls for awareness of 'planetary boundaries', including global water resources. The Human Development Report (UNDP, 2020) warningly adds that in order to re-stabilise planetary systems, human development has to be renewed, mainly by inserting technology and innovation especially into developing countries. In this 'Technocene' (Lemmens & Hui, 2017), human needs and environmental sustainability depend on the successful implementation of technology (Dube et al., 2020). Nevertheless, technological fixes alone will not suffice, and humanity—mostly the Global North—has to re-evaluate its relationship with the environment and 'relax its grip on nature' (UNDP, 2020: 3; see also Castree, 2015; Stubblefield, 2018).

Against this backdrop, one could easily be disturbed by this paradox: Who controls whom, then? Water humankind or vice versa? Social scientists and human geographers tend to shed light on a third variant, namely how some powerful human actors control water and therefore control people's access to water. The Anthropocene is criticised as an Anthropo(Obs)cene (Swyngedouw, 2019) or Capitalocene (Moore, 2016) because it is on the one hand depoliticising associated global environmental changes and on the other built 'upon flawed conceptions of nature, history, and humanity, rending it an impotent construct from which to respond to ecological crises' (Stubblefield, 2018: 1). By depoliticising environmental disasters, powerful actors, organisations, institutions and their actions disappear behind the 'human' (Gebhardt, 2016), dissociating the causers from their effect. 'Techno-managerial dispositifs' have evolved as the central pillar of environmentalism, which hides the political agenda of these actors, mainly the wealthy and powerful, who intend to retain their power (Swyngedouw, 2019: 253). They promote environmental protection, on the one hand, in order to secure their profitable natural resources, on the other. Thereby, social injustice and inequity are actively reinforced while these constructed 'natural' disasters are blamed upon *all* actors: 'Ecology is the new opium for the masses' (Badiou, 2008), as it controls—or rather paralyses—through the fear of an imminent dystopia and therefore vindicates the commitment of technological solutions despite its possibly negative outcomes. Techno-managerial solutions essentially enable change, 'so that nothing really has to change' (Swyngedouw, 2019: 254). This Marxist point of view basically tries to convey the idea that 'it is easier to imagine the end of the world than to imagine the end of capitalism' (Jameson, 2003: 76). Moreover, rather than overcoming the antagonistic dualism between humanity and nature and then challenging or changing capitalism, humanity will largely stay on the 'business-as-usual' trajectory while trying to mitigate its catastrophic socio-ecological effects until planetary boundaries are stretched to such an extent that only '(re)shaping' the environment through geoengineering will preserve humanity (Stubblefield, 2018). Here, the primary contribution of social sciences is that they shift the perspective from 'natural' to 'human-made' environmental disasters, while blaming specific actors rather than humanity as a whole.

In the passage above, however, the reader may have noticed that all of a sudden, the terms 'nature' and 'water' have been used synonymously. When I referred to nature, I included water as an inherent part thereof, while society apparently exists somewhat outside of 'mother nature'. This normative perspective nonetheless has been challenged and marks an important entry point for the following chapter. At first glance, simple, yet profound questions have to be asked in order to clarify why technocratic, modernist and capitalistic processes could inflict such damage on the environment and therefore on water and humans. What exactly is nature, what is society and—more importantly for this thesis—what is water? If the fundamental premise of our hydrological discourse—water itself—remains unaltered, the prospect of a water crisis will inevitably persist. It is, therefore, imperative to rethink and question the very nature of water (Linton, 2010).

This chapter's purpose is to give an understanding of the core principles of the so-called 'hydro-social cycle', which has developed over the last two, almost three decades in the political ecology of water.

The main goal is to understand that according to the proponents of the hydro-social cycle, water crises are the result of not only a physical lack of water, but also an ontological fallacy; that is to say, the modern understanding and management of water is based on a complex historical process in the hydrologic sciences as well as in the political and technological endeavours of the so-called 'hydraulic mission' that has run since the twentieth century. At the end of this chapter, I wish to convey that next to re-evaluating the human-water relationship, it is also mandatory to elucidate the 'operators' of this relationship (Latour, 2017: 17–18), namely the scientists—and other actors—who produce, discuss, distribute and sometimes manipulate knowledge about human-water relationships. A deep dive into (academic) socio-technical discourses will allow scientists studying the hydro-social cycle to unfold numerous *controversies* and, as such, different concepts of water. Therefore, a landscape of a variety of actors and their *water ontologies* is revealed.

2.1 Water(s) and H_2O—Sciences and Science

Probably most current encyclopaedias define water simultaneously as H_2O (its essence) and as the basis of life (its function)—a juxtaposition with which common people would probably agree. Water is H_2O, and H_2O is water. But is this tautological? Are they indeed both the same? The former used to be a mystical element, a divine material or in some religions a divinity itself. The latter appears to be the hard kernel, a debunking, demystifying fact of what water really is—an object discovered by rational modernity; not holy, not spiritual and not divine. As such, the signifier 'water' has devolved into an empty word that floats by different names in different languages around the well-defined and universally accepted chemical compound H_2O. It is a single truth, surrounded by 'less accurate' and 'less real' socially constructed terms. H_2O, in terms of precision, is the very substance of water, while its ability to provide life is inscribed into this chemical compound. If one accepts this definition of water, however, a gap opens up between theory and empiricism, because H_2O abstracts water into a homogenous monolith despite its heterogeneity. To clarify this point, I use a simple example of water distillation, where I assume the formula 'water = H_2O'.

In chemistry, distillation describes the separation of H_2O from adventitious substances such as minerals. The terminology distinguishes water containing a high level of minerals, referred to as 'hard water', while the opposite is classified as 'soft water'. If water equals H_2O, then soft water can only be regarded as 'real' water, while the other can be categorised as an 'impure' or a 'lesser' water. The translation of natural scientific fact into reality generates a few issues. In the following, we see an example of how fuzzy some expressions turn out to be:

> Water for [...] preservation of health has driven man to devise methods for maintaining its purity and wholesomeness. The water can get contaminated, polluted and become a potential hazard to human health. *Water in its purest form devoid of natural minerals* can also be the other end of spectrum where health could be adversely affected. (Verma & Kushwaha et al., 2014: 377, emphasis added)

There is water that is contaminated (unhealthy), water that is 'pure and wholesome' (healthy) and water entirely purified of anything else but water (unhealthy). All three forms have something in common, i.e., H_2O, but only one of them is designated as 'water in its purest form'. Consequently, only water that is entirely constituted by H_2O *and nothing else* equals the essence of water; other forms are therefore deviated versions of 'pure' water. The issue here is that one accepts such phrases perhaps too quickly, without realising the political notion of such a seemingly scientific statement. Indeed, the process of defining water is political, in that any definition is a 'circle' that distinguishes those who belong within the circle and those who do not. To be clear, the author above could have, more accurately, written 'Water devoid of natural minerals is H_2O'. So why was it necessary to imply a concept of 'pureness of form' that resonates rather with religious concepts than a simple scientific description? In fact, 'purity of water' stems from spiritual or religious practices whereby, in Judaism or Islam, for example, people undergo washing or bathing rituals to 'purify the believers both physically and spiritually' (Tvedt & Oestigaard, 2011: 21). Water cleansed of its impurities is H_2O. Water that is not uncleansed is H_2O plus adventitious substances. The political nature of such a statement is the implication of what water is (H_2O) and what it is not (anything attached to it)—a clean line that separates the purity of water from its contaminants. In this case, there persists an idea that there must be some*thing* capable of drawing the smallest ontological circle around water; or, put differently, only one, pure ontology of water that evades social constructions entirely (Fig. 2.1). The discussion on what water is has come to an end, since H_2O offers a way to reduce it to its core. As the Oxford Learner's Dictionaries (2024) writes, 'Water is the name given to the chemical compound of oxygen and hydrogen with the chemical symbol H_2O'. However, is this by scientific standards an accurate depiction of water?

The scientific purifying ritual of water that is H_2O applies only to one place: laboratories. In the common world, water occurs in solutions,

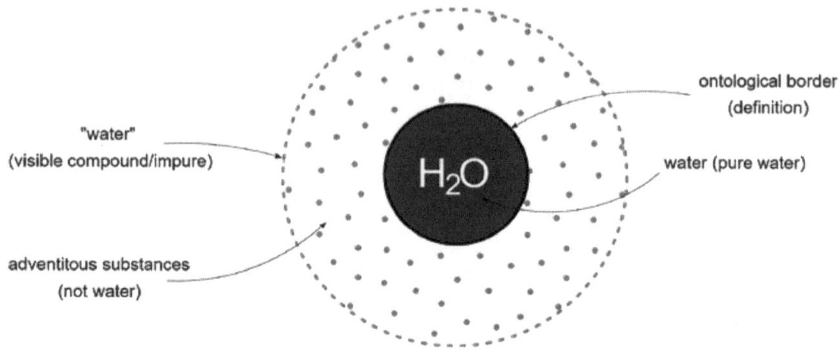

Fig. 2.1 The definition of water according to Science

i.e., 'homogenous mixtures distinguished from single substances called compounds' (Needham, 2000: 20), and consists not only of H_2O. As Linton (2010: 4) further notes, 'Even H_2O [...] is shockingly promiscuous – it goes and bonds with practically everything once it escapes the lab'. So far, the baseline may sound like this: water has a common chemical compound (H_2O) and everything attached to it. Alas, though, its very nature still has not been captured, because its very foundation, H_2O, remains uncertain.

Chang (2014: 248), in his book 'Is Water H_2O?', concludes that the scientific claim might be true, but nonetheless 'we can't just call any bunch of H_2O molecules thrown together "water," without doing violence to up-to-date science'. According to recent studies, water might be more than one complex liquid with 66 properties and rather 'two simple liquids with a complicated relationship' (Brazil, 2020; see also Gallo et al., 2016)—a statement which may appropriately describe the relation between humankind and water: simple, but complicated. Water is simultaneously a resource 'upon whose constancy (of both quality and quantity) we depend' (Bakker, 2012: 617), for which there is no substitute and which is an unco-operative commodity (Bakker, 2003). Consequently, any exclusive attempt to manage or regulate it is hampered by its multifunctionality (Barnes & Alatout, 2012; Linton, 2010) and its reluctance to fit into the confines of human concepts. The main takeaway at this point is not to defame natural sciences, scrutinise scientific and infest objective knowledge with an unnecessary objection of social constructivism but to explain that even such certainties like 'water

= H_2O' can be questioned, in which case the modern understanding of water itself can be left open to further investigation. Addressing or defining water in a natural scientific manner is not an issue; however, thinking that it is the only right way, is so. This has less to do with a relativist position and more to do with an awareness that there are multiple ways to refer to—and thus define—water, as will be shown herein. What should be avoided is what Latour (2013: 98) calls the 'idealism of materialism', which defines the lived world entirely according to a crossing of scientific findings and the world that has been analysed. Again, the intention is not to call the scientific mode of natural sciences invalid, but hamper the modernist urge to call this mode as hegemonic, capable to define everything universally above all other definitions—including water.

The other takeaway is that it is mandatory to stress the difference between the sciences and Science, intentionally spelled with a capital 'S'. Latour (2004: 249) clarifies and specifies this difference, in that the former implicates a politicised version of the sciences, which wrongfully presents nature as an already unified entity that threatens public life. Conversely, sciences emphasise the plurality of entities and realities, expressed through the term 'pluriverse' (Latour, 2004: 246). The main argument provided by Latour is that neither should the sciences nor nature be perceived and presented as singular, monolithic institutions. Simplified, yet scientifically incorrect statements such as 'water = H_2O' are based on the abstract idea of sciences, namely Science. While sciences—not only recently—have argued that more matter belongs to water than only H_2O (Liao & Schmidt, 2023: 7), the politicisation of sciences (Science) has distorted the perception of water based on seemingly scientific facts.

Therefore, it seems that H_2O must not equal water, and water still evades the scientific endeavour to define it for good—at least for now. What water 'really' is, according to the physical and chemical sciences, is in fact knowledge, which will evade most ordinary people, as it is constantly moving and 'discovering' new characteristics thereof. If we accept that water can—but need not—be defined as H_2O, then the question arises as to whether there are alternative ways to define it. In contrast to natural sciences, where water is pinpointed as a mere *thing*, in social sciences it is often addressed in a more holistic, inclusive manner, which renders it more than just matter.

Blurred Lines While natural sciences try to define water by its ingredients on an atomic level, social scientists such as the political ecologist and human geographer Erik Swyngedouw wish to open up its ontology. Their main argument is that water blurs the line between the dichotomous worldview of nature, the sphere of objects and society, the sphere of humans. Moreover, it ignores this line and renders it obsolete, which changes the constitution of water itself. As such, Swyngedouw's definition echoes this rebellious notion:

> "Water" is indeed not just H_2O; its meanings and practices meander like rivers, making unexpected turns and gathering or assembling all manner of connections and relations, transforming the social and physical landscape. (Swyngedouw, 2015: 20)

In this case, water is similarly matter and an immaterial, interacting entity comprised of different meanings and practices, all of which form its environment accordingly. However, according to this 'definition', water goes well beyond its materiality and merges with the social or rather human 'sphere'. It is neither a pure natural object nor exclusively a social actor, but both, a hybrid. Nonetheless, such perspectives on water's 'real nature' seem to overemphasise its social aspects and turn it into a rather elusive and mystifying entity. Conversely, the geographer and environmental historian Terje Tvedt insists that 'water is H_2O in nature just as it is H_2O in society' (Tvedt, 2021: 8). Therefore, he wishes to maintain a dichotomous ontology, where water is his own agent nonchalantly traversing the world of humans and nature. However, despite the chemical reductionism of water, Tvedt and Oestigaard (2011: 6) also extend its ontology well beyond its material properties, thereby resembling Swyngedouw's perspective:

> [...] water is the same in both nature and society and is also both a natural object and a natural bond, as well as a social bond and a discourse.

In general, this society–nature squabbling is either at the forefront or silently in the background of water ontologies. Hastrup and Hastrup (2015: 7) fittingly comment on this issue:

> Another question is how water may "oscillate" between nature and culture; this probably has less to do with water than with the human mind, seeking to understand the world in such a dualistic manner. Even while trying to dissolve the opposition between nature and culture, it resurfaces in

> statements that seem convincing at first sight, until one realizes that the argument is still steeped in a dualistic worldview that waterworlds explode.

There is a variety of perspectives on water's ontology and its designated ability to blur the line between nature and humankind. From the perspective of the hydro-social cycle, however, which will be further elucidated below, the main issue of current water crises also stems from its strict ontological separation from society, as it is put into the natural sphere. If this were indeed true, then the very 'nature' of water could only be defined and addressed by sciences (primarily natural sciences)—an issue that was discussed, for example, by Latour in 'Politics of Nature' (2004).

This dichotomous split between nature and society was also elaborated Latour (2004) when he used Plato's allegory of the 'Cave', whereby humankind (society) exists in a cave, and nature (reality) is set outside. Humankind is a captive of its limited perception and believes that the shadows cast from the outside (nature) on the cave's wall *are* reality. Only the scientists are capable of escaping the cave and returning to it to shut down any discussions on reality; a scientist thus becomes a 'heroic figure, that of the Philosopher-Scientist, at once Lawgiver and Savior' (Latour, 2004: 11). Latour, however, criticises this allegory and with it the idea that the world is divided into society and nature (2004: 16):

> Is this how people really live? It hardly matters. The myth requires first of all that we humans descend into the Cave, cut our countless ties with reality, lose all contact with our fellows, abandon the work of the sciences, and begin to become uncultivated, hate-filled, paralyzed, and gorged with fiction. Then and only then will Science come to save us.

To a certain extent, this allegory also applies to water, because if it remains outside of society, and thus in nature, it inevitably turns into an apolitical, ahistorical object that can be 'made seen' only by scientists and experts as they reveal its 'real' nature liberated of any distorting social explanation. As such, these scientists and experts are the only ones who can show the 'primitive cavemen' how to think of and use water properly. Lay expertise would turn into a nuisance, as has already been the case (Venturini & Munk, 2022: 89–91). In this cave allegory presented by Latour, Science would have a monopoly on reality, and politics would profit from its ontological work, as we have seen especially for western nations in the last century. Thus, science and its production of knowledge, as well as politics, are deeply intertwined (Seguin & Lord, 2023: 35). H_2O, detached from the social, cultural sphere represents a hard fact, a singular truth, and regardless of whether H_2O is water or not, the conviction persists in the modern discourse that water can be reduced to a simple fact. The purification of all waters is a symptom of modernity, a wish to cleanse natural objects from 'dirty' social attachments. Latour criticises this notion for at least two reasons. In his book 'We Have Never Been Modern', Latour (1993: 38) uses the expression that the 'moderns' speak with a forked tongue. While these moderns pride themselves on their rational, scientific and profound development projects, liberated of cultural and religious influences, they arrogantly diffuse any other approach as 'irrational', 'backwards' and 'archaic'. When Latour targets the moderns, he is in fact attacking and scrutinising the slogan, the conviction of modernisation—not the moderns in a geographical sense—which suggests a constant forward movement, a modernisation front that leaves behind a past that was bluntly speaking worse in all aspects (Latour, 2018). However, even H_2O and Western concepts on water, such as the water cycle, can be historicised and thereby its social traces disclosed.

Modern Water's Origin This was the main task of Linton's book 'What is Water?' in which he argues that water has become a modern abstraction, one that can be managed as a 'particular kind of water, stripped of its complex social relationships such that it may be managed by experts who are not necessarily directly involved in these relationships' (Linton, 2010: 58). The mainly natural science-focused perspective on water has been an 'epistemological revolution' which has successively blindsided modern societies to water's social aspects. This is exemplified and/or discussed by numerous scholars, such as Swyngedouw (2015) in 'Liquid Power', Strang (2004) in 'The Meaning of Water', in the three volumes of Tvedt et al. (2006) on the 'History of Water' and Hastrup and Hastrup (2015) in 'Waterworlds'. Even a growing number of natural scientists insist on intensifying the collaboration between hydrologic and social sciences in order to reveal the complex relations entangled in water resource management (Montanari et al., 2015; Gober & Wheater, 2015; Sivapalan, 2015; Xu et al., 2018). Notably, while these—and many other—scholars challenge the presumably technocratic perspective,

> [...] few scholarly works have been published about how water has been conceptualised and perceived at different times and in different societies

> although all social systems have a hydraulic dimension and water has been interwoven with social interaction from profane activities to religious ceremonies all over the world from time immemorial (Tvedt & Oestigaard, 2011: 1).

For example, the dichotomy of water (nature) and society (humans) that has been imposed epistemologically by modern science is the result of a long-term, historical process. In Europe, the conceptual gap between humans and water widened due to religious influences, and before the rise of Christianity, especially in the medieval period, there existed across early Europe

> [...] a rather "organic" relationship between humankind and the material world. Without the objectifying distance provided by scientific deconstruction and technological advancement, there prevailed [...] a more intimate relationship with the material environment, in which the properties and processes of other species and things were recognized as having their own agency. (Strang, 2015: 253)

As such, non-humans (in this context, water beings and deities) were an integral part of humankind and therefore of society. Over time, however, new anthropocentric religions and established technologies transformed this egalitarian order via a new hierarchy that put humankind on top of the pile. The more water was controlled and scientifically demystified, the more grew the feeling of dominance over it. This tendency increased especially with the rise of monotheistic religions. For example, in Britain, after the Norman Conquest in 1066, water expressed and symbolised the political control of powerful elites, while increasing scientific understanding prompted abbeys to promote it as 'the gift of God and part of a hydrotheological cycle of water movement according to his plan' (Strang, 2015: 260). The material dominion of humankind over nature, aligned with theological influences, initialised the development of Linton's (2010) modern water.

2.2 Modern Water: From Atoms and Cycles to Hegemony

The starting point of what we view as 'modern water' varies, from the 1800s to 1850s (Hamlin, 2000: 321) to the mid-seventeenth century (Linton, 2010: 77). Whichever period is considered, though, two major events were at the breakthrough in Western civilisation, namely the Scientific and the Industrial Revolutions, both of which catalysed 'ideas about human dominion over and distinction from nature' (Strang, 2015: 268) and sustainably changed the perspective on water and, ultimately, the water paradigm.

Hamlin (2000: 321) argues that modern understanding has shifted from 'waters' to 'water', which arguably emphasises the point made above by Latour, in that waters the many (pluriverse) have turned into a homogenous mass (nature/universe) through the politicisation of sciences (Science). This paradigm shift meant that i) water was demystified as an element, ii) its potential to diffuse infectious diseases was highlighted and iii) it was recognised as a public necessity to be universally provided with 'the maximal standard of purity' (Hamlin, 2000: 322). Although H_2O did play a part in this paradigm shift, Hamlin (2000: 321) warns not to overemphasise this scientific finding, since other factors were also relevant (e.g., industrialisation, consumer concerns, changes in philosophy and religious beliefs).

The history of H_2O proves that the construction of the compound was gradual and scientifically contested during its roughly century-long genesis. In the 1770s, Joseph Priestley, A. L. Lavoisier and others showed water to be a compound of hydrogen and oxygen. From the eighteenth to the nineteenth century, H_2O was successively formed: at first, Dalton illustrated water figuratively as atoms and molecules, Berzelius proved water to be in a 2: 1 hydrogen- and oxygen-atom ratio and it was finally condensed by Thénard into H_2O (Linton, 2010: 78–80; Chang, 2014). Yet, this scientific finding did not close the debate on water up until the 1860s. Moreover, the Chemical Revolution was unable to disprove other systems that regarded it as an element. Chang (2014: 204) goes so far to state that

> [...] even at the consensus-point reached after half a century of atomic chemistry, there was no absolute proof of that formula H_2O. On the whole, a century of chemistry [...] failed to provide sufficient reason to

close the debate on the constitution of water. Yet, the debate did close, and "Water is H_2O" became an apparently unshakable scientific fact.

This, however, does not imply that H_2O would be a wrong conclusion or unscientific, but it somehow reached a hegemonic state and the status of an undeniable fact that nowadays violates up-to-date science, as discussed above.

The scientific findings collectively created a narrative that tried to glorify its achievements while simultaneously despising a seemingly archaic, premodern past. Modernity was eager to draw a sharp line between itself and the 'unenlightened' era beforehand. In contrast, earlier concepts of water were misrepresented, implying that it was mostly uninteresting and only 'an abstract issue of interest [...] to natural philosophers and metaphysicians' (Hamlin, 2000: 314). Around the same time, this idea of the purity and impurity of H_2O was intensified.

The narrative of societies' detachment from water was pushed even further with the cholera epidemic in Great Britain. In 1855, John Snow discovered in London that water and not miasma—as scholars usually claimed—contaminated cholera, which ultimately led to a large-scale construction of proper sanitary services. Therefore, water's pathogenic potential became a major criterion to classify it as either safe or unsafe. Yet, this implicitness was built on a presumed ignorance prior to modern society. In the nineteenth century, the public relations campaigns of sanitarians led to the conclusion that, previous to Snow's revolutionary recognition, people were insensitive to their environment and hygiene, i.e., water was something incidental in their day-to-day lives and nothing to worry about (Hamlin, 2000: 314–315).

This ignorance was even transferred to non-Western societies, which, for example, legitimised colonial powers to dispose others of their water—since, from the colonisers' point of view, 'modern' state and scientific rationality were deliberating in contrast to other obsolete alternatives (Mosse, 2008). The attitude of modern water culture has not changed in its essence, as Linton (2010: 74) critically notes that '[water] is disclosed to us primarily through modern scientific practice – we have learned to know water in the way that we do through scientific discourse and modes of representation'. Linton (2010: 80) goes even further with his point of criticism by incidentally summarising what 'modern water' is:

> In scientific chemistry, the nature of water was defined as some *thing* to which all waters could be reduced – H_2O – a basic chemical compound that drove out all the socially specific qualities to different waters. The shift from premodern waters to modern water was the product of an epistemological revolution; it occurred as a consequence of a particular way of knowing and representing water in modern and scientific practice – a practice that presumed a fundamental separation between the natural and the social.

Besides the fact that water was both reduced to a chemical compound and its human actors dissociated from it, it was pushed exclusively into the natural sphere by the imaginary of the hydrologic cycle. This is the second main argument made by the hydro-social cycle.

The Genesis of the Hydrologic Cycle Hydrology can be defined as 'the science that deals with the processes governing the depletion and replenishment of the water resources of the land areas of the Earth, and treats the various phases of the hydrologic cycle' (Rosbjerg & Rodda, 2019: 109) or, as James Dooge (1983: 23) simply phrased it: 'Hydrology [centres] on the occurrence and movement of water on our planet'. As these definitions illustrate, hydrology is primarily concerned with the water cycle, therefore making it its very foundation. However, the hydrologic cycle, as the following will show, has very differentiable genealogies, much less linear and, above all, less ancient than presented in modern hydrology—so goes the argument of the hydro-social cycle. There exist at least three histories relating to how the hydrologic cycle was discovered or constructed, namely the history proposed by modern hydrologists, a more critical version, suggesting a rather political and pragmatic notion, and one that discloses religious influence.

According to the hydrologic sciences, the development of the hydrologic cycle was a successive development dating back to ancient times (Marcinek, 2007). Although the water cycle has been philosophically described in ancient China and Middle Eastern civilisations, and more prominently by Aristotle and Marcus Vitruvius, the scientific approach to hydrology failed to appear up until the sixteenth century (Rosbjerg & Rodda, 2019: 110). It was firmly believed that '[…] water *rose* from the sea or from the earth in underground channels to provide the fresh water that we see on earth's surface' (Linton, 2010: 111). During the Renaissance, at the beginning of the sixteenth century, Leonardo da Vinci set the starting point for a scientific approach to water. He had multiple interests in this element, for example the water cycle and flood protection, for which he proposed solutions such as canalisations, damming, building locks and a new gate for the Arno River in Italy (Pfister et al., 2009: 1399). In 1575, Bernard Palissy observed that, indeed, springs originated from rain, and in 1674, Pierre Perrault claimed also that rivers could be supported by rainfall—an assertion verified by Edme Mariotte in 1686. Around 1700, Edmond Halley added evaporation to the equation, and finally, in 1802, John Dalton completed the water cycle, thereby laying the foundations for modern hydrology (Rosbjerg & Rodda, 2019: 110). This narrative of hydrology draws a linear transformation from natural philosophies to natural sciences and is meant to encourage practicing scientists to deliver them '[...] with a sense of continuity by knowing their professional roots' (Back, 1963:ix; quoted in Linton, 2010: 110). Yet, although the genesis of the hydrologic cycle is almost exclusively associated with this story, which Linton (2010: 109) describes as the 'scientific hydrologic cycle', there exists another version.

In his book 'The Hydrologic Cycle and the Wisdom of God', Yi-Fu Tuan (1968) discloses how the modern depiction of hydrology is abbreviated and disentangles the influence of Christianity. Natural theology, especially in England, contributed to conceptualising the hydrologic cycle from ca. 1700–1850. Among them, John Ray (1627–1705) used scientific findings made at that time to offer evidence that the cyclical movement of water had to be an architectural design of God (Tuan, 1968)—and in 'God's plan' there was plenty of water for dry lands, as the next excerpt from Ray indicates. In a response to critics, who saw no sense in why God would create such a large sea and so little dry land, Ray (1962: 10; cited in Linton, 2010: 121) wrote, 'The Wise Creator [...] did so prudently order [water], that the sea should be large enough to supply Vapours sufficient for all the Land [...]'. This quotation represents what Tuan identified as a northern temperate bias of the hydrologic cycle, where surface water is copiously available. Although this 'sacred hydrologic cycle' (Linton, 2010: 108) was discredited by 1900, due to the growing influence of Darwinism, the imagination of water's generous and abundant circulation remained. Take, for example, the explanation from Marcinek (2007: 33; emphasis added):

> The hydrologic cycle is the main phenomenon of hydrology. Only water, mostly drinking water, will be renewed in this cycle very *quick* and *continuous. Gigantic quantities* of water are present, from which the hydrologic cycle takes the water.

Thereby, a dominant, mainly Western idea evolved, which Linton (2010: 123) framed as 'hydrologic Orientalism', whereby deserts and arid lands were regarded as barren, poor and uncivilised. Therefore, hydraulic re-engineering was seen as a solution, a part of God's plan carried out on several continents by his colonial or neo-colonial apprentices in the form of modern water.

While modern hydrology roots itself in natural philosophy and the Scientific Revolution, Linton (2010) objects to this genesis. In his opinion, hydrology and the hydrologic cycle were founded and created by the American Robert E. Horton in 1931. Of course, by the 1900s, hydrology existed in some form, but it struggled as a scientific discipline. First, it was mostly seen—at least in America—only as the study of underground water phenomena, due to the fact that, second, the science of meteorology already occupied large portions of the water cycle by investigating 'evaporation and run-off', 'river and flood work' and 'mountain snowfall and water supply investigations' (Linton, 2010: 134). Therefore, hydrology's claim for the atmospheric sciences was regarded as somewhat redundant in the scientific community. Second, hydrology has the reputation for being primarily driven by 'practical and technical demands' instead of a 'disinterested pursuit of knowledge about the behaviour of water' (Linton, 2010: 136). However, in 1931, Robert Horton managed to embed hydrology as an engineering science and has been declared as the father of US-American hydrology (Linton, 2010: 143). In this context, Linton (2014: 112) claims that Horton's illustration of the water cycle consolidated hydrology's existence as part of the geosciences by giving it an identity, thus demarcating its scientific territory and equating the hydrologic cycle to hydrology. With the 'invention' of the 'Hortonian Hydrologic Cycle', hydrologists launched a historical retro-perspective to identify renowned philosophers and scientists as the forefathers of the hydrologic cycle. The variety of the first advocates range from Homer's Iliad, Marcus Vitruvius, Leonardo da Vinci, Theophrastus, Xenophanes, Anaximander and Anaxagoras. In this context, Linton (2010: 146) critically notes, 'With respect to the variety of these attributions, projecting the hydrologic cycle backwards into the writings of people for whom it did not exist is bound to be a somewhat haphazard affair'.

To conclude, each of the three versions of the hydrologic cycle reveals a certain loophole. The scientific hydrologic cycle seems to be constructed and invented by modern hydrologists in order to provide the discipline with a sense of history and identity, the sacred hydrologic distorted the image of the water cycle as being abundant and sufficient for all the dry land as part of God's plan and, finally, the Hortonian Hydrologic Cycle apparently served as a political instrument for American hydrologists to justify and cement their raison d'être. This is not to say that a hydrologic cycle does not exist, but as Linton (2010) points out multiple times, it is not an apolitical, ahistorical or natural concept but is infused with religious concepts, idealism and politics, rather than decoupled from it in a—often propagated—dispassionate manner.

Moreover, presenting the hydrologic cycle as the sole reality of water's movement risks misrepresenting the diverse ways in which water is found in different geographical places and human settlements. It is a model, a template that tries to give a sense of general validity, but it is derived from the climatically humid parts of the world, translated into every other part of the world, which renders them abnormal. Furthermore, the hydrologic cycle divides water from society by putting it into the realm of nature. This imagination creates an antagonistic relationship whereby people can only access water by altering the hydrologic cycle (Linton, 2010). Lastly, the cycle metaphor creates an image of flow that is both constant and seamless. Sudden changes like floods or stagnant water raise questions about 'the stability of our cyclical representations of water' and whether this 'metaphor of cycle also excludes other forms of hydrologic and ecological processes that do not behave according to patterns of circulation and repetition' (Camargo & Cortesi, 2019: 5).

The hydrologic cycle therefore has a point 'at which the story began', to quote Strang (2015: 273), who further proposes a much more entangled relationship between humans and water:

> The material properties of water continue to flow into the human imagination, carrying ideas about its powers: its capacities to generate life; its potential to carve landscapes and transform itself and other things over time; its ability to subsume events and to disappear. The deep influence of being immersed in these material events is a contestant in human lives, continually stimulating, in one cultural form or another, ideas about the

agentive powers of water and nature itself. Ever greater religious and technological hubris on the part of humankind has failed to subsume or completely control these powers; one might say, instead, that it has merely raised the level of the dam.

2.3 Modern Water: A Material and Ontological Crisis

Strang's metaphor of an ever-growing dam describes the modern relationship between humans and water, especially since the twentieth century. For 150 years, the Western world firmly believed it 'had a tight grip' on water. Proto-scientists, politicians and hydraulic engineers successfully turned the dominant epistemology of water as a—devoid of its social nature—mere natural resource into a dominant ontology (Swyngedouw, 2004), i.e., 'Western societies saw themselves as the only conscious agents in a passive material environment' (Lenton & Latour, 2018: 1066). The modern perspective on water, which began in Western Europe during the mid-nineteenth century, diffused throughout the world during the imperial reign (Mosse, 2008).

Non-modern Hydro-Societies Nevertheless, the control of water proved to be less universal and axiomatic than supposed by the colonisers. An ideal example to prove the collision of worldviews on water is the imperial period in European history during a time of global expansion, especially in terms of the British Empire and India. In the Indian Province Bihar, south of Nepal, rural inhabitants adapted their lives around the major alluvial rivers that flow into the Ganges, by 'going with the flow' of the river and relocating their settlements, on the one hand, to avoid severe flooding and on the other hand to profit from fertilising river sediments. In the nineteenth century, British rulers imposed their water management system in the region despite not understanding local relations between water, ecology, society and the economy (Claus et al., 2015: 296). They assumed that the construction of flood-controlling embankments would stabilise the market value of land as a commodity (Claus et al., 2015: 296). However, these implementations had the opposite effect and even aggravated floods. And despite the fact that the British recognised their mistake and removed their constructions, technocratic, modern water management still prevailed in the region, even after India's' independence. Nowadays, the Bihar region has become synonymous with floods, as the thousands of kilometres of embankments cause destructive deluges that wash away villages and towns (Cortesi, 2021; D'Souza, 2016).

Another case where modern water and traditional management approaches met, occurred during British rule in Egypt. This encounter might even explain why Ethiopia's dam project on the Blue Nile is now causing turmoil and political instability in the region (Gonda, 2023). Egypt's dominance in the Nile basin was primarily shaped by the colonial rule of the British Empire that single-mindedly sought to anchor its influence in Africa along the course of the Nile. Tvedt (2021: 49–60) explains the high monetary investment in Egypt's water infrastructure by the colonial rulers as follows. First, it allowed irrigated agriculture to expand in order to grow profitable cotton for further processing in British textile mills. Second, a constant water supply guaranteed the pacification of the otherwise seditious Egyptian population, which was predominantly engaged in agriculture. Third, control of the Nile meant an economic and military strengthening of Egypt, so that the British Empire had less need to use its own resources to sustain its position of power in Africa. On the one hand, this hydraulic mission involved the development of structures such as dams, canals and dikes; on the other hand, they were also a means to an end, territorially manifesting the establishment of a British Nile Empire stretching from Lake Victoria to the Mediterranean. Driven by this task, the British decisively shaped the discourse around the Nile, and as a result, water was declared scarce for ten million Egyptians in 1893. Even if the vision of the British Empire failed, the exploitation of the Nile for capitalist purposes, and the political-military superiority along the course of the river, possibly haunt Egypt to the present day (Gonda, 2023).

As these two examples show, the narrative of controlling water via idealised modern technologies and sciences, while prioritising political and economic interests, began during the rule of imperialist powers such as the British Empire. Moreover, despite the downfall of western imperial reign during the twentieth century, modern water fostered its hegemony—and most successfully so in the USA.

The Impossible Mission From the mid-1890s onwards, the US-American Republic party pushed their agenda of controlling water on a national scale to make profit from what were, at that point, untamed natural resources (Wester et al., 2009: 396)—a phenomenon termed the 'hydraulic mission' (Worster, 1985). Consequently, water was stripped from private enterprises and placed in the hands of the state (Bakker,

2002), or in other words, water went from a local to a centralised state affair led by 'civil engineers staffing hydraulic bureaucracies (hydrocracies)' (Molle et al., 2009: 328) 'bent on "developing" water resources' (Wester, 2009: 10). The expansion of American territory to the west, alongside economic success and military power, resulted from this hydraulic mission or, as Linton (2010: 161) expresses it in hydro-social terms, 'The state […] materialised modern water, while modern water helped build the state in a kind of reciprocal ratcheting process that contributed to making America the foremost economic and military power in the latter part of the [20th] century'.

The epiphany of the hydraulic mission was clearly symbolised by large-scale dams[1] Especially, from the 1950s to the 1980s, the majority of the globally 61,988 registered large dams were constructed (ICOLD, 2023; Perera et al., 2021: 7). Flyvbjerg et al. (2017) summarised the dams of the twentieth century as a 'Trojan Horse' masked as being for purely economic purposes but operated as political and ideological instruments of public governance and financing organisations to level all 'unruly communities' (2017: 14). This levelling was—and maybe still is—the mantra of the modernising hydraulic mission: 'We shall all be equally modern, all equally capable of profiting from what, alone, forever escapes the tyranny of social interests: economic rationality, scientific truth, technological efficiency' (Latour, 1993: 131). Thus, the modernising front (Latour, 2018: 29) moved on with the hydraulic mission as the new modus operandi of water resource management.

The first cracks in this techno-managerial foundation appeared in the 1950s, as one water crisis followed the other (Linton, 2010: 200–203). Thereupon, a Neo-Malthusian conviction, propelled by politicians, media and a global network of hydrologists, declared the world's water resources to be scarce and conveyed and imaginary of finiteness—somewhat opposed the abundancy of the hydrological cycle. Additionally, the proliferation of modern water management technologies caused sustainable societal, ecological and economic losses. In the USA, for example, the total capital investment for water over the twentieth century amounted to about 400 billion US$ (Gleick, 2000: 128). The once so successful centralised and state governed hydrocracies were overloaded by their own

[1] Large-scale dams are defined as: "a dam with a height of 15 metres or greater from lowest foun-dation to crest or a dam between 5 and 15 metres impounding more than 3 million cubic metres" (ICOLD, 2011: 3).

ambition to control water, but even the diversion of costs through the privatisation of water since the 1990s has proven to be a failed experiment (Gleick, 2018). The concept of social evolution, the idea that modernity would finally seal the victory over the unruliness of nature, was neglected by Tvedt and Oestigaard (2011: 7), who insisted that 'the more humans seek to control water and socialise it, the more power water and variations in the water landscape will exercise over societies'. Certainly, powerful actors have controlled and instrumentalised water since the very beginning of human history, but there is a power inherent in water that has made modern societies so dependent—and therefore vulnerable—to water stress and/or scarcity.

The need for clean drinking water represents the very dilemma of modern water management. Surface water, especially in developing countries, is polluted by open defecation. Consequently, the UN (2015) and collaborators have promoted the use of groundwater for drinking purposes. Despite the success of hand pumps reliably accessing shallow aquifers, the proliferation of this technology has had severe consequences, as shallow aquifers have been proven to be contaminated with arsenic. In Bangladesh, for example, millions of people were contaminated over long periods of time (Sultana, 2007) while the primary focus on groundwater led to the depletion of regional aquifers, resulting in both water scarcity and tensions between rural and urban communities, followed by water grabbing by urban areas (Birkenholtz, 2016). Although the ocean appears to be an uncontested, ubiquitous and almost perfect source for freshwater desalination, it does not solve economic, social or ecological issues but further exacerbates them. By reverse engineering the hydrologic cycle, desalination plants produce fresh water abundantly from inexhaustible seawaters, thus fulfilling the increasing gap in demand in the Middle East and North Africa regions, where 48% of the worldwide 15,906 desalination plants are situated. Although this technology seems promising, increasing water scarcity and decreasing water quality result in relatively high economic costs and environmental issues (Jones et al., 2019). The Arabian Gulf countries, for example, depend on desalinated water to satisfy both their population and an ever-growing economy, but the increase in desalination increases environmental damage. The process, among others, pollutes coastal waters with microplastics, chemicals and metals, contributes to poor air quality by creating a photochemical smog (Paleologos et al., 2018) and increases sea water temperature by 7–8 °C in the vicinity of desalination plants (Bleninger, 2010; Dawoud,

2012). Certainly, the technology may improve in the future, but reversing its recent consequences could be as complicated as decarbonising the atmosphere to mitigate anthropogenic climate change.

A (Post-)Modern Paradigm Shift The need for a paradigm shift in today's water management is self-evident. The question remains, however, in which direction this transition should move. So far, the general consensus is that modernity in the twentieth century was marked by water policies that relied on the hydraulic mission, which proved to be beneficial for billions of people but was also unsustainable from a social, economic and ecological perspective (Gleick, 2003: 1524). Allan (2005) argues that the Global North has left the trajectory of the hydraulic mission and entered a new trajectory of reflexive modernity, especially since the increasing advent of environmental movements.

To overcome the challenges of climate change, and hence water scarcity, a 'soft path for water' or 'integrated water resource management' are proposed as sustainable alternatives. The goal of this transition is to complement current water management systems with decentralised systems, 'open decision-making, water markets and equitable pricing, application of efficient technology, and environmental protection' (Gleick, 2003: 1524).

At the same time, this transition might be hampered by certain barriers such as

> [...] overcoming tradition and culture, antiquated laws and institutions, inertia in complex social systems, the long time required for changes in technology, inadequate financial investment, and more. Sometimes, individuals or groups with an interest in maintaining the status quo hold far more authority or power than those with an interest in implementing new approaches. (Gleick, 2018: 8869)

Paramount to the success of this paradigm shift is still economic growth. According to Gleick (2018: 8868), economic growth and sustainable water management will decrease water consumption, as seen in the USA and Singapore, for example. The underlying hypothesis guiding this argument is the so-called 'environmental Kuznets curve'—something to which even Gleick (2018: 8868) refers. The premise is that if a country reaches a certain threshold of wealth, environmental challenges can be dealt with accordingly. However, there is a contrasting perspective on this notion, as established by Damania et al. (2020): the more an economy grows and develops, the more it intervenes with its waterscapes and leaves behind its footprint. Highly developed countries have reached a point nowadays where water has a 'quality unknown', as termed by Damania et al. (2020). They argue that a country cannot grow out of this problem economically, since 'pollutants of poverty' (usually easier to detect and trace) disappear in highly developed countries yet are substituted by 'pollutants of growing prosperity & emerging pollutants' (difficult to identify and trace)—increasing in range and number annually (Damania et al., 2020: 3). The environmental Kuznets curve and its devotees believe that economic development decreases humanity's pressure on the environment, but reality proves that highly developed countries tend to interfere even more with the environment in myriad—sometimes untraceable—ways.

Although a paradigm shift is indeed needed, the proponents of the hydro-social cycle argue that in quasi-sustainable solutions, modern water is still the ontological basis; thus, no real change is possible. This can be exemplified with the European Water Framework Directive (WFD) and the Integrated Water Resource Management System (IWRM).

The main directive of the Water Framework Directive is environmental protection. The objective aims to achieve 'good ecological status or good ecological potential and good chemical status for surface waters, as well as good quantitative and good chemical status for groundwater by 2015' (European Commission, 2019: 9). Even though the deadline has been extended to 2027, the goal set initially was far too ambitious. Indeed, the European Commission (2019: 1) acknowledges the fact that:

> no substantial progress in water bodies' overall status has been made between the first and the second river basin management cycles. The Directive's implementation has been significantly delayed and less than half of the EU's water bodies are in good status, even though the deadline for achieving this was 2015.

The Water Framework Directive, however, blames external factors for its failure, such as lack of data, financial resources or the implementation of objectives in other EU policies (European Commission, 2019: 1). This goal, however, as Linton and Krueger (2020) emphasise, could never be achieved due to its 'ontological fallacy'. They argue that the WFD a priori is unable to achieve its goals, but even more so, it 'constitutes part of the problem; that is to say, the construction of this system is itself based on a shaky (ontological) foundation that renders it unworkable in the real world' (Linton & Krueger, 2020: 516; Melo Zurita et al., 2015). During the modernisation phase, or the Hydraulic mission, in the twentieth century, waterscapes could be transformed by human ingenuity, driven by hidden political and economic interests, with minimal consideration for the social and human implications (Latour, 1993). Latour (2018: 41) points out that the failure of modernisation could be registered as soon as 'the place "on" or "in" which we are located begins to react to our actions, turns against us, encloses us, dominates us, demands something of us along in its path. How are we to distinguish from now on between physical geography and human geography?' But so does the Water Framework Directive by emphasising environmental protection that displays ecocentrism in the foreground and yet is run by a neoliberal ideology in the background (Linton & Krüger, 2020: 517). This neoliberal orientation has every characteristic of biopolitics, which strives to optimise the environmental conditions without discouraging economic growth (Fernandez, 2014). Therefore, the WFD's model 'helps [to] legitimise a view of environmental protection and improvement that is fully consistent with conventional models of economic growth' (Linton & Krueger, 2020: 521). Besides the neoliberal imprint of the Water Framework Directive, the claim that they strive for a pristine nature is in itself deeply anthropocentric. The waterbodies' 'reference conditions are far from naturally given; rather, they are outcomes of complex negotiations between technical provisions and conceptual premises' (Linton & Krueger, 2020: 524). Environmental protection as such has nothing natural about it. A similar criticism has been voiced by scientists towards the IWRM, described by the Global Water Partnership (GWP, 2000: 22) as

> [...] a process which promotes the coordinated development and management of water, land and related resources, in order to maximize the resultant economic and social welfare in an equitable manner without compromising the sustainability of vital ecosystems.

A seemingly democratic attempt offered by the IWRM is the inclusion of all relevant stakeholders that as such partake in the decision-making process. Yet again, both ontological and political issues have been disclosed by empirical and theoretical studies. Ontologically, the stakeholders gathered in the process can express water only in terms of modern water. IWRM is still *resource management* (Linton, 2010: 241), and therefore water can be addressed only technologically or as a hydrologic scientific fact. Therefore, other associations of water are neglected, and hence stakeholders are excluded and the modern dispositive of water management hegemonised: 'The beneficiaries of management decisions according to critics are not individual people any more seen as collective political subjects, but rather the population as a mass, which are the victim of external environmental problems' (Graefe, 2016: 47). This notion depoliticises water and proposes techno-managerial solutions as saviours despite the fact that other voices become silenced. For example, Mehta et al. (2014) highlight a multitude of concerns with IWRM in South Africa, Mozambique and Zimbabwe. In summary, these systems fail due to (i) a lack of institutional co-operation with local communities, causing subsequent mistrust and opposition among water users, (ii) emphasising power asymmetries, (iii) exclusion of stakeholders, (iv) discriminatory water allocation, (v) privatisation and pricing of waters by (foreign) investors, who have no cultural relation to the region, and vi) a lack of understanding in terms of women's water needs and demands, as well as their exclusion from decision-making processes. In summary:

> [T]he case studies demonstrate both inconsistencies in implementation, and what the depoliticized policy narrative of IWRM ends up obscuring [;] namely, reasserting the power of the state and its ability to facilitate the allocation of water to powerful users and render some users more visible than more marginal ones. (Mehta et al., 2014: 539)

Similarly, Clement et al. (2017) observed such tendencies in Nepal, whereby IWRM is presented as a holy grail despite its failure to register water as a political and social resource that is very much capable of producing social injustice. IWRM transforms waterscapes into valuable assets for economically valuable actors, while other actors can no longer associate themselves with the new neoliberal water regime. In this case, water actually does achieve 'blue gold' status—as claimed Neo-Malthusian—because it shifts from a public good to an almost exclusively private good controlled by 'choreographed [...] market mechanisms' (Swyngedouw, 2009: 58). Moreover, more industries, agriculture and urban areas intervene with other non-human actors, which makes environmental protection even less probable, let alone any ambition to return to a state of environment without people. Alas, the environment's only contribution in this scenario is to guarantee economic growth.

Therefore, despite the ambitious projects proposed by the WFD or IWRM, these systems have nothing but unveiled their economic nature, which cannot help but instead reproduces modern water and surrounds itself with powerful actors who intensify and distribute the modern water dispositive. Less powerful actors who do not—or simply cannot—participate in this ethic of growth (Gleick, 2000: 128) experience a marginalised existence, or perhaps worse, they either lose a sense of identity, leave their homes or in the worst case are unable to react to their new surroundings and are therefore left paralysed in what is for them an increasingly foreign environment.

2.4 Water as an Actor: The Hydro-Social Cycle

It is at this point that the Political Ecology of Water has developed increasing interest in the dynamics of water privatisation and marginalisation and how social, political and economic actors have co-produced this allegedly apolitical, techno-managerial environment—co-produced by water. Water has to be politicised, as it flows and intervenes in human society, while powerful interests cause asymmetrical access to it –in terms of both quality and quantity—and thus socio-environmental issues (Swyngedouw, 2006; Wilson et al., 2019), which in turn characterises water as a social actor[2] and not as a mere, inanimate object. In this sense, an actor describes something that has agency, whereby it is 'the tension that makes the actor' and not the intentionality or force (Latour, 2017: 53; 57; 66). Furthermore, the agency of 'things' even becomes self-evident in the anthropomorphising language that is often used when civil engineers, for example, address the actions of a river, which resonates a rather aggressive vocabulary. A river 'gives', 'supplies', 'captures', 'prevents' and so on (Latour, 2017: 53–54). Similarly, when different agencies overlap, combine or intervene, a vast and complex association of actors unfolds, if one, for example, considers the development of a water reservoir:

[2] According to Latour (1996: 7), 'An "actor" in AT is a semiotic definition – an actant–, that is, something that acts or to which activity is granted by others. It implies no special motivation of human individual actors, nor of humans in general. An actant can literally be anything provided it is granted to be the source of an action'.

> The water held in place behind dams [...] is not merely the liquid H_2O measured in cubic metres that falls through penstocks and turbines to generate hydroelectricity; this water is held in place by state-run-power utilities, the human labour that is extracted to produce the dams, penstocks, and turbines; abstract hydrologic calculations; water management protocols, discourses linking national identity with generation of hydroelectricity; networks of transmission wires; consumer expectations; construction consortiums; and political discourses; which together have the effect of fixing it in a particular way. This water is what it is by virtue of events and *processes* that transcend the place and time of the water itself. (Linton, 2010: 30–31; emphasis added)

The dam and the water reservoir are arguably 'power tools', which in turn help actors to enforce their dominion (Meehan, 2014); moreover, as Linton emphasises, these tools are shaped and maintained by an array of diverse human and non-human actors. The focus on processes, which make it likely that water *has been made* to be stored behind a dam for certain purposes, has some resemblance with Latour's (2013: 32) argument that a gas pipeline is not made of gas but of a 'heterogeneous set of elements that allow circulation to occur'. This network, or better, 'work-net' (Venturini, 2010: 266), of actors entangled in the process of developing, for example, water reservoirs, water pipelines, canals and desalination plants, marks an important gateway to the hydro-social cycle—a concept that has been developed by human geographers since the 1990s.

Although the term has become popular since Linton and Budds' (2014) article 'The hydro-social cycle: Defining and mobilizing a relational-dialectical approach to water', it was especially Swyngedouw's work in the late 1980s and 1990s on urban water studies that laid the foundation for the hydro-social cycle (Larsimont & Martin, 2022: 106). Swyngedouw drew on a diverse set of theories constituting the core ideas behind the hydro-social cycle (Larsimont & Martin, 2022: 106): Marx's metabolism, Lefebvre's production of space, Harvey's relational dialectics, Latour's concept of hybrids and Haraway's cyborgs. All of these concepts converged into the hydro-social cycle in order to overcome the binary of society/culture and nature/objects and to emphasise the relevance of politics and neoliberal markets in the manipulation of water's movement, which eventually led to the memorable statement 'water flows to money' (Linton & Budds, 2014: 173).

In the following, I elaborate on the core principles outlined above and present a number of papers that have contributed empirical case studies to the hydro-social cycle. Moreover, I disclose the pathway I have chosen to hopefully contribute some empirical insights from Lake Urmia for the hydro-social cycle, and perhaps its further development. Since this inquiry focuses on an empirical case study and not on a theoretical development of the hydro-social cycle, I have limited my theoretical contribution to the discussion section at the end of the thesis.

Water-Human Relationship In contrast to the ontological reductionism of water to its chemical composition, non-modern water ontologies, such as that of First Nation people in Ontario (eastern Canada), expresses the specific interrelatedness of humans and water. Water is synonymous with a living, animated entity, capable of feelings, which is considered sacred and 'not a single, discrete aspect of the environment [,] [but] part of a greater, interconnected whole', to which access should be granted to all entities in need of water, including flora and fauna. Moreover, water is addressed by some elders as a relative, which anthropomorphises it and yet moves it closer to people's lives on an ontological level; water is to be treated with dignity and respect (McGregor, 2012: 10–11). The hydraulic mission, as mentioned above, was the polar opposite of this rather intimate, perhaps mystifying portrayal of water. Water, in the course of the twentieth century, has become a mere resource—if one is convinced by the arguments of the hydro-social cycle's proponents. Latour (1993: 50) argues that the modern division of nature and humans, and the management of natural resources, 'was remote and under control', as long as 'nature seemed to be held in reserve, transcendent, inexhaustible, distant enough'. However, in the face of the global water crisis, it seems that water has been reanimated and is striking back at humanity—both 'moderns' and those on the way to becoming 'moderns'—in the same way that Gaia, 'Mother Earth', has done in the context of the climate crisis (Latour, 2017). The relationship between water and humanity is one that has developed together throughout history. This suggests that any ontology that separates these two concepts as two distinctly different, natural and cultural entities is an obsolete perspective. Thus, in order to understand the nature of the water crisis, one must engage with actors who declare it as such and identify those who allegedly cause the crisis. This has been one of the main tasks of the hydro-social cycle: to highlight the relationship between humans and water, rather than predisposing their (antagonistic) opposition.

Evidently, a quasi-modern dichotomy of humans and nature does not correlate with the hydro-social cycle. The theory therefore leans on contemporary 'Western science [that] is underpinned by a vision of all beings – including humans – as part of a universal nature' (Strang, 2015: 253). Water cannot be externalised solely to the sphere of nature: it is at the same time part of a 'hydro-social landscape' (Swyngedouw, 2015: 21). Against this backdrop, Orlove and Caton (2010: 402) suggest that water should be studied as a 'total social fact'—in the sense of Mauss (1990: 80), who sought to identify 'the social "things" themselves, in concrete form and as they are.' First, the dichotomy of nature and humans must be overcome, which the hydro-social cycle does—in its estimation—by referring to socio-natures as everything there is. In theory, this ontology makes it impossible to detach the social from the natural, and vice versa (Castree & Braun, 2001: 13). Moreover, the hydro-social expands the nature of things by adding relational dialectics to highlight the reciprocal processes between humans and water.

The term 'relational', combined with dialectics, originates from Harvey (2010). The claim that water is more than H_2O and just a natural object only solidifies when the relation between things is not regarded as fixed, separated and superficial. On the contrary, things as of animate and inanimate 'objects' constitute each other, or in other words, they are what they are due to their relationship, but consequently, they are interwoven and therefore inseparable. Social, political, cultural, spiritual and scientific processes are reflected in the way in which water is treated (Swyngedouw, 2015: 21), changing both water and human society in a 'spiral of challenge-response-challenge' (Worster, 1985: 22), where both humans and water have an agential role, as both 'frame particular social worlds' (Hastrup, 2013: 59). This 'dialectic depends on contexts and interconnections to provide the essence of reality' (Chatterjee & Ahmed, 2019: 374), or, to offer an artistic metaphor, objects in an artwork are put in relation to each other; this is considered composition. While each object would be just a fairly complex 'thing' in itself, as a whole artwork they communicate, evoke emotion and transport meaning to the viewer according to the colour pallet, the positioning of objects, geometry, symmetry, asymmetry, the expression of figures, light, shadow and so on. Similarly, the hydro-social cycle does not focus on one specific subject but on the process—or more specifically the dialectics—of water, human society and technology (Linton & Budds, 2014: 176), i.e., the endless 'dialogues' offered by theses, antitheses and syntheses between these three entities, so to speak (Fig. 2.2). Linton (2010: 41) argues that 'the process itself' is foundational, and 'the need to look behind or below it to find a particular driving or determining force' is avoidable. The artwork example demonstrates the modus operandi offered by dialectics, because instead of starting with an object and then connecting the other parts 'to reconstruct the larger whole', dialectics initiates with 'the whole, the system or as much as of it as one understands, and then proceeds to an examination of the part to see where it fits and how it functions, leading eventually to a fuller understanding of the whole from which one begun' (Ollman, 2003: 14). Furthermore, dialectics opposes the analytical tradition in the sense that the latter 'involves the disaggregation of the whole into discrete component parts that are then defined in isolation as fragmented aspects of a totality' (Sayers, 1991: 142). As such, dialectics and Lefebvre's (1991) 'ontological priority of processes and flux' (Swyngedouw, 2015: 21) are consistent.

Water and Power Geography in particular has elaborated the elusive and complex network of relationships amid 'people, places and ideas',

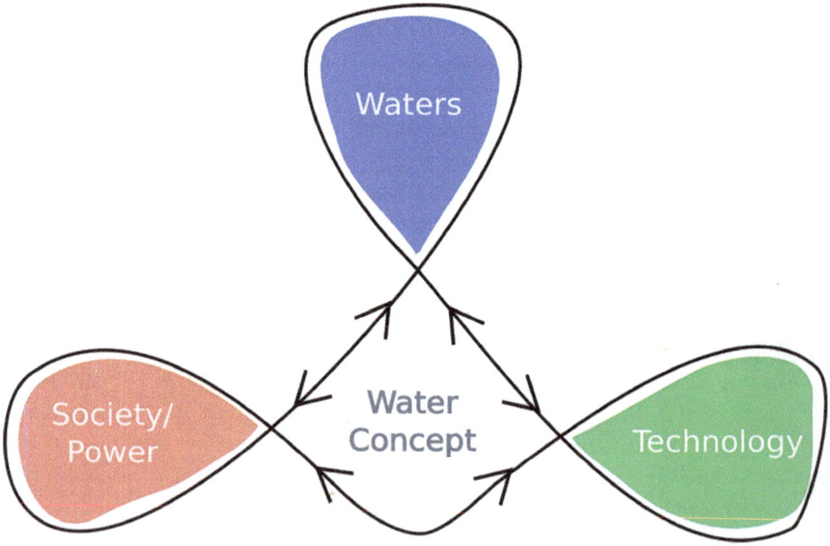

Fig. 2.2 Hydro-Social Cycle; own adaptation based on Linton and Budds (2014: 176)

which are decisive factors in the occupation, distribution and dispossession of water resources (Menga & Swyngedouw, 2018: 3). These power struggles over water, despite gaining prominence in recent decades, have been analysed for roughly 70 years.

The relationship between water and social power was first theorised in Karl August Wittfogel's seminal work 'Oriental despotism: A comparative study of total power' (1957), which built on Marx's dialectical interpretation of the labour process to express how humanity simultaneously manipulates nature to satisfy its needs and thereby changes its own nature (Linton & Budds, 2014: 173). Wittfogel applied this approach to water to elaborate on the dialectical relationship between large-scale irrigation systems and centralised political authority in arid and semi-arid regions of ancient hydraulic civilisations such as China, Egypt, Mesopotamia and India. In essence, this theory explained how social elites manifested their power through hydraulic infrastructure, created big bureaucracies and concentrated their political power that Wittfogel classified as 'despotic'. Water was regarded as a central component in the creation of a centralised state power, and it was transformed by society and transformational for society in equal measure (Linton, 2010; Linton & Budds, 2014; Obertreis et al., 2016). Wittfogel's thesis, however, was criticised for its technological determinism, empirical selectiveness and attention to a limited number of countries (Worster, 1985). Nevertheless, his work was pioneering in terms of the further analysis of the relationship between nature, technology and society.

The research on power relations in human-water or social-nature systems has gained momentum since the mid-1990s (Castree, 1996, 2001; Swyngedouw, 1996, 1999), and for about two decades the 'hydro-social cycle' concept has gained growing attention. Through relational dialectics, hydro-social scientists aim to show that, indeed, things like nature and water have a history and therefore can be historicised (Swyngedouw, 2015; Linton, 2010). Furthermore, water resources are both the product and producer of history (Mosse, 2008). The concept of the hydro-social cycle has emerged from critical or radical, human geography grounded in Marxist theory, hence the employment of historical and materialist analyses (Wesselink et al., 2017). The literature on anthropology, which asserts a notion of water and people in an experiential, cultural and metaphorical sense (Hastrup, 2013; Hastrup & Hastrup, 2015; Krause & Strang, 2016; Mosse, 2003, 2008; Orlove & Caton, 2010; Strang, 2004), certainly influenced the theory of the hydro-social cycle, just as much as philosophers such as Latour and Whitehead did. Consequently, the hydro-social cycle applies a variety of theories, as elaborated so far. Its peculiarities can be explained by its distinct difference in relation to socio-hydrology, for example (see Wesselink et al., 2017) in terms of ontology, perspective and scale. The latter two are a consequence of the hydro-social cycle's origins in political ecology, where researchers evaluate the interrelations between and socio-ecological influences of decision-making processes from a large scale (e.g., international, national) down to a small scale (e.g., individual land managers) (Robbins, 2012). Swyngedouw (2015: 21) expresses this view as follows: 'If I were to capture some water in a cup and excavate the networks that brought it there, (now citing Latour (1993: 121)) I would pass with continuity from the local to the global, from the human to the non-human'. Although the hydro-social cycle and socio-hydrology recognise water and society's close relation, the latter calculates the human factor quantitatively through settlement patterns and economic prosperity (Wesselink et al., 2017), while the former focuses on social power expressed through the implementation of water infrastructure and technology, i.e., 'the articulation of water and social power relations' (Linton & Budds, 2014: 173). Swyngedouw (2009: 58) in this context points out that:

> Water research has for too long concentrated on either the physical side or the managerial side of the water problematic, often tiptoeing around the vexed question of how political economic power relations fuse the physical and the managerial together in particular and invariably socially uneven ways.

Boelens (2013) furthermore criticises the overwhelming amount of modernistic, technical–economic water development prescriptions and universalistic modernising water laws in most intervention programs instead of recognising and investigating the so-called 'smallholders'—those who are meant to adapt the government's applications for water usage (e.g., farmers). These quotes show how the hydro-social cycle is concerned with both the lived consequence on the field and the power of politics and the discourse on water.

The ontological difference between the hydro-social cycle and socio-hydrology lies in the perception of water. As Hastrup (2013: 60; emphasis added) states, 'water *does* something in society', yet by this she means more than 'just' making life possible. Water, an active agent and hybrid, is both a product and producer of socio-nature, as 'it carries people's thoughts' (Hastrup, 2013: 60) while being transformed by them and transforming society—as explained through relational dialectics. On these terms, Swyngedouw (2004; 2015) tries to overcome the nature-society dichotomy of water by re-introducing Marx's concept—similar to Wittfogel—of metabolism, according to which humanity changes matter through a process which is 'organized under distinct social and political arrangements' (Swyngedouw, 2015: 27) and therefore actively produces socio-nature. Political projects embody a process of socio-environmental transformation, while every socio-environmental project reflects and materialises a particular political vision (Swyngedouw, 2015). Hydro-social scientists are thus driven to analyse the processes by viewing the hydro-social landscape as

> an assemblage of interwoven processes that are simultaneously human, non-human, material, discursive, mechanical, and organic, but ultimately driven by political forces and economic processes that aspire to turning nature into capital, a process that necessarily implies changing social relations to nature. (Swyngedouw, 2015: 21)

Moreover, people engage with non-human nature in discourses, asserting that 'ideas can play an important role in the production of water' and 'ideas [...] [also] have power' (Linton, 2010: 26). This, in part, is why besides the Marxist approach to water-related power studies, a Foucauldian approach to the hydro-social cycle has gained some attention (Boelens et al., 2016; Rattu & Véron, 2016). Instead of limiting an analysis of society through a 'capitalocentric' (Gabriel, 2014) lens, power could be diversely spread as a vast network of micro-powers (Rattu & Véron, 2016) or 'power relations between people, as opposed to a quantum possessed by people' (Kelly, 2009: 37 quoted in Balan, 2010: 38). A discourse analysis elucidates how power and knowledge maintain 'a specific political order, as if it were a naturalised system, by making fixed linkages and logical relations among a specified set of actors, objects, categories and concepts that define the nature of problems as well as the solutions to overcome them' (Boelens et al., 2016: 7). Yet, those who are governed may oppose the 'naturalizing discourse' by using 'tactics of noncompliance and everyday acts of resistance' (Scott, 1985 in Benjaminsen & Svarstad, 2019: 393).

In recent years, the hydro-social cycle has been applied in numerous case studies to elucidate the politics of water (Paerregaard, 2018; Workman, 2019). On a global scale, Wiegleb and Bruns (2018) criticise the UN's SDG number six 'Clean Water and Sanitation' for three reasons. First, a paradigm shift to sustainable water management, or a soft path, seems unlikely, as the economic and financial system overall still dictate the outcomes of water management projects. Second, social and environmental issues are treated separately rather than being holistically linked to the water cycle. Third, the co-operation of multiple stakeholders runs the risk of focusing on individual targets instead of emphasising the overall objective. Apart from demanding a less managerial approach to water, Wiegleb and Bruns (2018: 1164) suggest 'acknowledg[ing] the political nature of water.' Multiple case studies reveal how modern water, although proposing economic welfare and sustainability, fosters its hegemony politically.

Workman (2019) elaborated how the implementation of IWRM in Lesotho caused water scarcity due to the political rearrangement of water management. Although water management was meant to be decentralised and driven by local communities and chieftaincy, Lesotho's government undermined local authorities and covered this approach as apolitical, modern and community-driven. While the IWRM proposes water security and equity, conflicts over water management responsibility have increased. The collapse of water management committees hence creates local water scarcity (Workman, 2019).

Millington (2018) explains how in Sao Paolo Brazil peripheral residents were more impacted by water scarcity than propagated by politics. The city's water infrastructure, as well as urban governance and varying capabilities of residents to store water, co-produced an asymmetrical water crisis.

Large-scale water management projects, which aim to improve water supply and water quality, tend to ignore social inequities. This was exemplified by Laituri (2020) in rapidly growing urban areas in Kenya, Indonesia and Ethiopia. She concluded that these technical, engineering solutions do not address social issues adequately, which simultaneously leads to a lack of community involvement for these large-scale projects.

Paerregaard (2018, 2019) focused on the water ontology of traditional water management in the Peruvian Andes and its clashing with modernist endeavours to introduce new management systems. In this context, Paerregaard (2018) concluded that water has three forms of power: one physical, one social/political and the other cultural and imaginary.

Flaminio (2021) investigated the clash between different water ontologies in large-scale dam-projects, namely hegemonic modern water and marginalised non-modern waters. The main argument of her paper is that conflicts revolving around these projects result in socio-technical controversies. By elucidating these controversial topics and the different parties involved, one may identify different water ontologies; that is to say, conflicting understandings on water, which in theory and practice are incommensurable, thus cause the controversies. Flaminio (2021) builds her approach, among others, on Yates et al., and's (2017: 809) argument that it is mandatory to identify different water ontologies, next to the hegemonic ontology of Western science—or Science, as I have argued above. In this sense, however, alternative waters must not be perceived as mere cultural constructions but as a lived and accepted reality for certain actors. In other words, as a researcher, one should address alternative waters with ontological dignity. Socio-technical controversies and the production as well as the transformation of knowledge mark an entry point for the inquiry, as suggested by Science and Technology Studies.

2.5 Insoluble Theories—Reflections and Discussion on the Hydro-Social Cycle

There lies some beauty in complexity. Not for the sake of being complex, but in the way it occasionally works at the front-end: simplified and—at best—catchy. The complexity, meanwhile, works in the background, everything conducted in such a way that only the artist, musician, craftsperson, athlete or researcher knows how much effort it cost him or her to reach seeming effortlessness. Yet, what if simplification covers up a *compromised* complexity, full of contradiction and paradoxes, or an oversimplification hinders access to its underlying complexity. I know that I might idealise my personal taste in terms of what defines complexity as 'good' or 'bad' from a recipient's point of view, but it is still an important question to ask: What makes scientific research easy to access without losing its complexity and experiencing degradation to popular sciences? Not asking such a question leads quickly to the other end of this spectrum, where a Gordian knot of theories and excessive rhetoric cause more disturbance than clarity. The hydro-social cycle finds itself in this category in many ways despite its effort to simplify a complex issue. Wesselink et al. (2017: 7) state:

> [W]hile its advocates assert that "the hydro-social cycle offers a critical approach that prompts us to consider how water internalizes and reflects social and power relations that might otherwise remain invisible," it is not always clear how the case descriptions would be different if less advanced theoretical weaponry had been deployed. Put differently: we contend that asking the simple question "how did the situation become as it is, and why?" with an open mind to all possible factors, including one's own biases, would yield the same rich and critical case descriptions without the need for sophisticated theorising.

Even worse, the 'theoretical weaponry' presented on the one hand overshadows the miniscule, concrete tools or methods on the other: 'It is often not clear how this theoretical positioning actually affects the case description and analysis, especially since most of these papers are very brief on the methodology employed' (Wesselink et al., 2017: 7). This can be exemplified through the vague analytical tools Swyngedouw (2015: 23) described in his book on the hydro-politics of Franco's Spain in the twentieth century. These six points essentially try to (i) reconstruct the processes to gauge 'transformative knowledge about water and the waterscape'; (ii) prioritise ontologically and epistemologically the 'process of becoming and of hybridisation'; (iii) contend that 'anything' can serve as an entry point for the study; (iv) state that 'this archaeology,' already begun, will never end and therefore is 'always open, contested and contestable'; (v) note that 'knowledge, discourse, and practise are invariably situated within the web of social power relations that defines and produces socio-nature' and (vi) proscribe the idea that 'socio-natural production transcends the binary distinctions between society/nature, material/ideological and real/discursive.' Except for the glimpses that suggest a discourse analysis and this 'archaeological' or rather historical process reconstruction, no tools are offered here. Instead, Swyngedouw emphasises how the hydro-social cycle changes a political ecologist's state of mind, therefore opening up multiple entry points for the analysis but remaining opaque in terms of its methods. Readers are expected to 'believe the interpretation provided by the authors, and it is not clear whether the researched subjects would agree with the analyses' (Wesselink et al., 2017: 7).

Linton's (2010: 241–247) proposed analytical hydrolectics, which offers a more sincere and less rhetorically overloaded tool in comparison to Swyngedouw. This might be due to the fact that Linton and Budd's version of the hydro-social cycle is still concerned with water-related issues and marginalised groups, but more from an ontological perspective than a historical-materialist approach. Hydrolectics asks researchers to 'identify actors involved' and associate them with different waters to understand what people *think* water is and what sort of implications their ideas on water discursively and materially might have (Linton, 2010: 242). Yet again, hydrolectics may answer the 'Why' and 'What' to do, but not 'How' to do it. It thus goes without saying that a two-step-approach to assess difficult tasks such as local water crises seems too simplistic and quite obscure in the face of the over-bloated theory presented beforehand.

Indeed, the hydro-social cycle tries to follow the threads through the Gordian knot (Swyngedouw, 2004: 21)—or maze (Swyngedouw, 2015: 21–22)—therefore emphasising socio-nature's production process, nevertheless making it rather difficult to follow at times. In my estimate, the theory of the hydro-social cycle marks both a solution and a problem, in that it opens up new ways to address water-related issues, on the one hand, but on the other it lacks a certain stringency—and therefore accessibility. Ultimately, the dialectic approach is meant to provide an overview of the studied subject and not just guide one through a disturbing maze disturbed even more by '[dialectic's] loops and spirals and other complex acrobatic figures' (Latour, 1993: 55). The focus should be on elucidating the vast network of humans and non-humans, presenting the process more systematically and guiding rather than just being sucked up in the historical-material process—by chance feeling as lost as some readers might be.

Perhaps the main reason why the hydro-social cycle reads and feels cathartic yet headache-inducing is due to 'a rather eclectic choice of [different] concepts between which the links are sometimes not convincingly established, but only loosely collated according to the aspect treated' (Graefe, 2013: 2). This imbroglio of theories is at the heart of the issue, which must be addressed urgently if the hydro-social cycle wishes to be more than just a catchy term that is more interested in solving a metaphysical issue than providing practical implications for research. The two main theories that have experienced a forced relationship are relational dialectics, or better, relational Eco-Marxism (Castree, 2002) and the actor–network theory. Linton—as Loftus (2011: 188) argues—'sees no rigid distinction between them [which Castree (2002)] has sought to reconcile what he considers to be false antitheses between the two approaches, the wounds from earlier tussles remain'. Furthermore, Loftus (2011: 188) would have 'liked more on this debate within the book ["What is Water?"], although it is perhaps wise of Linton to ignore it'. At the risk of being unwise, I am willing to take the risk, even though I might stumble over this large, overgrown Gordian knot.

Power to All Actors As mentioned above, Castree (2002) tried to identify the possible coexistence of both ANT and relational Eco-Marxism. He declared—what he supposed ANT advocates would do—five sins created by social scientists: social-natural dichotomy, local–global dichotomy, asymmetry, impoverished concepts of actor and action and power geometries. I would like to start with the latter to show how Swyngedouw and Linton violate ANT principles to push their hydro-social cycle in a direction that favours relational Eco-Marxism and as such leaves ANT concepts behind as mere superficialities to somehow just combine the natural and the social in the fashion of contemporary Western philosophies. This is not to say that the hydro-social cycle would be wrong by any means, but after all, Lave (2015: 216) rightfully points out that 'Swyngedouw only dipped one toe or two into ANT waters,' marking perhaps less than a half-hearted attempt to combine these two very distinct theories. The provocative statement will be self-explanatory, because if we highlight the differences between the power concepts of ANT and the hydro-social cycle, almost all of the five 'sins' will follow.

Swyngedouw does not define power but shows his contradictory views on it. On the one hand, 'power geometries [...] choreograph access to and exclusion from water' while this 'social power [...] fuses together with water management' (Swyngedouw, 2009: 59). At first, he tries to decentralise power in a network, performing disparities in access to water, which resonates with ANT in the sense that power does not belong to humans alone but also to non-humans (this will need more explanation, provided later on). However, Swyngedouw already disclosed that his source of power is anthropocentric by referring to the web of social relations (Swyngedouw, 2015: 23). Other than ANT, which has a different understanding of 'social' or 'society', where humans *and* non-humans belong together in a collective society, the hydro-social cycle still considers the former to be on the centre stage of power, whereby nature is reduced to an effect of social or human power. In other words, water does what people do, or, as Linton (2010: 3) provocatively wrote, 'water is what we make of it.' As such, the attempts 'to decentre power are more rhetorical than actual' (Castree, 2002: 132). Swyngedouw (2006: 114) even criticises Latour for his apolitical notion of ANT, stating that 'he not only ignores question of power, but dismisses it radically as a fruitful entry into the excavation of socio-natural imbroglios' (Swyngedouw, 2006: 113). Therefore, Swyngedouw chooses to side more towards Haraway's ontological position, whereby he describes entities as an embodiment of relations, which is compatible with the hydro-social cycle and its production processes of relational-dialectical quasi-objects, or put in a simpler way, how things materialise conflict and propel a never-ending transformation of themselves and their environment. To this end, Swyngedouw, citing Zitouni (2004; In Swyngedouw, 2014: 14), nevertheless ignores the fact that for Latour *and* Haraway

> […] power differences and inequality can no longer be stated as such, as a departure point into the city but have to be explained through the many actions and relations between objects, humans and non humans. There is nothing behind any space or agent, only attachments aside of it that make it stronger or weaker, allow it to exist or lead it to perish.

In the face of this quote, it is surprising how much Swyngedouw seems to ignore its content, in that he refuses to let go of either anthropocentric 'power' or 'dialectics', even though Zitouni marks the latter as an unsuitable tool for urban ANT studies. However, Swyngedouw (2006: 114) convinces rhetorically at first, when he wishes to see the city a 'complex [of] political, social, economic, and cultural relations […] invariably fused with myriad configurations of power that saturate material practices, symbolic ordering, and imaginary (or imagined) visions'. His source of power is neither hybrid (human and non-human) nor dispersed within his network, as he defines the hydro-social landscape as being

> […] viewed as an assemblage of interwoven processes that are simultaneously human and non-human, material, discursive, mechanical, and organic, *but ultimately* driven by *political forces* and *economic processes* that aspire to turning nature into capital, a process that supports and maintains social life and sustains economic growth always combin[ing] society and nature in infinite ways; yet these hybrid socio-natural "things" are full of contradictions, tensions, and conflicts. (Swyngedouw, 2015: 21; emphasis added)

All of the power resides in politics and capitalism that make social—or, perhaps better, modern life—possible as it is today. Not only is it difficult to 'imagine Latour approving these ideas of network and hybridity' (Lave, 2015: 216), but more so, they are antithetical to ANT. Castree (2002: 135)—one could say—comes to the aid of the hydro-social cycle when he argues that some actants are more powerful and 'marshal the power of others', hence reproaching ANT for its egalitarian power distribution among humans and non-humans. Nevertheless, this is not necessarily the case with ANT. For this argument to be clear, we need to look at the already mentioned deviating understanding of 'social' or 'society' by Latour to elucidate why ANT does not refuse power altogether but its 'aether'-like conception by social scientists—something that lures in the 'background' or 'above', pulling all the strings and triggers.

First, against all the misconceptions of ANT, power is not neglected. As Venturini writes (2010: 262):

> We do not believe power to be a universal resource that can be spent freely in every social domain. According to ANT, we conceive power as a local property that derives from holding a specific position within a specific network. The extension of such "power" depends obviously on the extension of the network itself. A child disobeying her parents and a political party contesting the results of an election are both fighting for power. The extension of their fights, however, is different.

Latour (2005: 72) on that note emphasises the need to study 'objects', 'not only as being full-blown actors, but also as what *explains* the contrasted landscape we started with, the overarching powers of society, the huge asymmetries, the crushing exercise of power.' In contrast to Tvedt's (2021: 201) assertion that nature would not be of interest to Latour 'because it does not exist' and would not be considered as a variable to explain social development, the opposite is indeed true: ANT places the objects at the centre of attention and asks researchers to study them first 'and only later at the standardised social' (Latour, 2005: 235). Therefore, Camargo and Cortesi (2019: 2) argue that the hydro-social cycle over-emphasises human power and renders water altogether powerless, although floods are the very proof of its unruliness. In some way or another, Linton and Budds (2014: 176) addressed this beforehand by expressing water's materiality, which is capable of stabilising or disrupting society. Regardless, Camargo & Cortesi—the former working in an STS department—disapprove of the idea that 'water is what we make of it'. They insist on water's own agency, which might break any harmonious, cyclical interaction with society, because it is, more importantly, 'what we cannot make of it' (Camargo & Cortesi, 2019: 3). By default, the hydro-social cycle has an inability to recognise 'ecological processes and non-human interactions in wet environments' (Cortesi, 2021: 874). ANT's dedication to giving power to objects might be difficult to comprehend due to the knee-jerk reaction that things like water act by some self-driven motivation usually attributed to animate beings. Yet, while Latour does not intend to grant some anthropomorphic identity to objects, he does essentialise their ability to be social, to make a difference and to co-produce society.

The following description by Serres (1982: 225)—in my estimate—is one of the best explanations summarising quasi-objects or quasi-subjects, neither being one or the other but 'both natural in society and social in nature', to sarcastically quote Tvedt (2021: 201).

> A ball is not an ordinary object, for it is what it is only if a subject holds it. Over there, on the ground, it is nothing; it is stupid; it has no meaning, no function, and no value. Ball isn't played alone. Those who do, those who hog the ball, are bad players and are soon excluded from the game. They are said to be selfish. The collective game doesn't need persons, people out for themselves. Let us consider the one who holds it. If he makes it move around him, he is awkward, a bad player. The ball isn't there for the body; the exact contrary is true: the body is the object of the ball; the subject moves around this sun. Skill with the ball is recognized in the player who follows the ball and serves it instead of making it follow him and using it. It is the subject of the body, subject of bodies, and like a subject of subjects. Playing is nothing else but making oneself the attribute of the ball as a substance. The laws are written for it, defined relative to it, and we bend to these laws. Skill with the ball supposes a Ptolemaic revolution of which few theoreticians are capable, since they are accustomed to being subjects in a Copernican world where objects are slaves.

Following this thought, it becomes clear why ANT does need not attribute quasi-objects and quasi-subjects with power but acknowledges their power all along. Power for ANT is a relational 'achievement' between humans and non-humans that needs to be constantly performed for certain actors to stay in the powerful position they occupy in a network

(Venturini, 2010: 271). Latour (2005: 64) chooses the term 'associations' simply to mark his dissatisfaction with the usual use of 'social'" whereby he designates social not as a 'domain of reality or some particular item, but rather [as] the name of a movement, a displacement, a transformation, a translation, an enrolment'. This marks an undeniable incompatibility with the hydro-social cycle, because it overdoes its theorising on how to overcome or 'transcend' (Swnygedouw, 2015: 23) the subject/object, nature/society dichotomy through the social and political production of socio-natures. Latour insists that this divide has never existed, and he cannot help himself but ridicule dialectical reasoning in this matter in almost every book he has written. Although Swyngedouw combines ANT, Harvey's relational dialectics and Lefebvre's historical–geographical process to conjure a 'transcendental' theory, he summons more metaphysical conflicts than some revolutionary transcendence. Bear (2016: 3) says correctly that 'socio-nature cannot be viewed as an easy reconciliation of historical materialism with poststructuralism, but rather as an ongoing critical dialogue.'

No Power to Capitalism In her article 'Reassembling the Structural', Lave (2015) not only critically points out the weaknesses of combining ANT and political ecologies, which in her estimate is an 'intellectual incoherence', more so, she also concludes that political ecology should let go of ANT because it does not resonate with the emancipatory agenda against neoliberal capitalism. Thereby, she calls ANT out for its notion to equalise humans and non-humans, ignoring 'legacies of colonialism, global political economy, racism, and sexism that shape who is enrolled and what level of influence they are accorded within that network' (Lave, 2015: 218), as well as subjects that have much in common with a rational *homo oeconomicus*. In summary, Lave (2015: 220) concludes that the strengths of ANT and political ecology are just that: different strengths. Taken together, both approaches exemplify intellectual incoherence due to their entirely different premises (Lave, 2015: 220). Haraway, so Lave (2015) argues, forms a much better relation with political ecology because she does consider humans and non-humans as forming a relational mode of existence that insists on a historical reciprocal co-production—quite similar to the hydro-social cycle's ideas on relational dialectics. However, in these relations humans are clearly the dominant 'partner', that, in Lave's (2015: 221) opinion, recognises 'the power dynamics inherent in significant othering strikes as a much more productive basis for an emancipatory politics of the Anthropocene.' Therefore, political ecology would not lose anything if it finally let go of Latour's ANT.

Castree (2002: 132) similarly sees 'the encounter between ANT and green Marxism [as] evidently one-way and deconstructive.' His conclusion, nevertheless, contains a watered-down version of ANT, because its liberating views might be valuable to 'discover[ing] something new,' albeit he warns that 'many on the geographical left – even those who distance themselves from Marxian work – would not want to give up on the metatheoretical project of making visible capitalist processes and relations' (Castree, 2002: 132). This new version highlights the need to incorporate Marx's ideas on capital, labour value and 'accumulation for accumulation's sake' into ANT to identify different 'socio-natural-commodity networks', which are embedded in the world of human and non-human actors, while the diverse and myriad 'natural' entities are the source of value. Although both humans and non-humans can exercise power and agency, it is money—the source of power in capitalist societies—that enables certain human actors to conduct their will over others, who serve as an intermediary, i.e., an actor translating action without altering or transforming it—which one could address as a 'perfect conduit'. The so far structural explanation, however, is misleading because this network is in a state of constant change and must be constantly worked on or 'in other words, the "thing" we call capitalism is constituted by hundreds of thousands of [simultaneously global and local] commodity networks, which mix different people, machines, codes and artefacts in often unique ways, but towards the same ends' (Castree, 2002: 140). The capital-nature relation can be characterised as 'a *creative* destruction' (Castree, 2002: 141), albeit not implying dogmatically that every transformation must lead to environmental degradation. In this context, the normative or positive definitions of 'good' or 'bad' environments demand from 'politically responsible eco-Marxism to judge different commodity networks in context using cognitive and moral discrimination' (Castree, 2002: 141).

Again, it is difficult to see Latour agreeing on such a union of ANT and Eco-Marxism as he and his colleague Callon clearly distance themselves from the idea of capitalism (Latour & Callon, 2011: 19):

> [W]hat can be done against capitalism? First of all, one should not believe in it. Capitalism, in the singular, is nothing more than the formatting of

> all of the interactions under the unique and singular reign "of" capitalism. Yes, capitalism is our only horizon but, happily for us, it does not exist! Thus, the first rule of method, of politics and almost of morals, is the following categorical imperative: nothing in your actions, your thoughts, or your research should further the unification of capitalism or its existence as the only mechanism of its own development. This rule should allow us to get rid of a good number of liberal assertions as well as of Marxism's pernicious fascination with capitalism, the monster that was to be "dialectically" overthrown and annihilated—the ruins of such a folly are still among us.

In his lecture on the 'Affects of Capitalism' in 2014, Latour labelled capitalism as meaning

> [...] not a thing in the world, but a certain way of being affected when trying to think through this strange mixture of miseries and luxuries we encounter when trying to come to terms with the dizzying interplays of "goods" and "bads". Capitalism is a concept invented to help absorb this odd mixture of enthusiasm for the cornucopia of riches that has lifted billions of people out of abject poverty and the indignation, rage and fury in response to the miseries visited on billions of other people.

These statements obviously contradict the Eco-Marxian theory of the hydro-social cycle—or Castree's alternative approach. Nevertheless, for Latour and Callon (2011: 2), 'capitalisms' or markets do exist, which at first resonates somewhat with Castree's (2002) concept of socio-natural commodity networks. Economy, however, does not play such a vital role for ANT. In 'Modes of Existence', Latour (2013) assigns networks with different qualities, what is meant to address a long-term critique, saying that ANT obscures differences between different networks (Castree, 2002: 134). Hereby, Latour calls 'The Economy'—similar to Capitalism—a utopian place, imagined by moderns, unreal, uninhabitable and as such an illusion: there is 'nothing that could even serve as a frame for [the Economy] (Latour, 2013: 426). Economy or capitalism, in his view, cannot be seen as driving forces within a network, and so he essentialises it as commodity network, as partly suggested by Eco-Marxists such as Swyngedouw, because Latour and ANT do not attribute economy—the trade of goods—with such a high priority.

The clear differences between ANT and the hydro-social cycle, or any Marxian-motivated political ecology, seems to confirm Lave's (2015) conviction that both theories have different strengths, but put together they seem to be insoluble and conflicting, even with Castree's (2002) watered-down version of ANT.

Apart from the hushed 'inner' conflict of combining ANT with Eco-Marxism, overburdening theories and a lack of methods, the hydro-social cycle and its main concepts mark an issue for many authors. One of them points at the very core of the hydro-social cycle: the water concept or, put simpler, what people think water is and how this thought translates into action, constructing a hydro-social territory through technologies, discourse and so on. At this point, although I lengthily explained how water turned into H_2O, or rather how Linton (2010) declared it 'modern water', I find myself in a bit of a conundrum. How do I know if people's ideas on water, their ontology, correlate with their actions? Undoubtedly, water is mostly known in our times through the sciences, but must this translate to a reductionist view on water? Latour (2004: 225)—and I beg pardon to overly rely on him even at this point—expresses this doubt as follows:

> The social sciences—economics, sociology, anthropology, history, geography—have a much more useful role than that of defining, in the actors' place and most often against them, the forces that manipulate them without their knowledge. The actors do not know what they are doing, still less the sociologists. What manipulates the actors is unknown to everyone, including researchers in the social sciences.

No one knows what drives people, but the hydro-social cycle suggests knowing both what people think and, as such, what makes them act. Against this backdrop, Lave (2015: 218) pointed out that 'there is no necessary correlation between the ontologically radical and the politically radical.' Schmidt (2014), in his article 'Historicising the hydro-social cycle', provided an example of the absent correlation of thoughts on water and materially modernising water.

At the dawn of the twentieth century, water management in the USA experienced a drastic change—as already mentioned. One of the leading figures driving this change, William McGee, so Linton argues (2010: 150f), declared water a resource: 'Water became known to the state in a way that made it an object of calculation and subject to a particular kind of accounting and manipulation' (Linton, 2010: 151). The reductionist view on water enabled the American state to populate the western area and thrive economically. Water or nature in general were nothing more but reservoirs for human needs. Conversely, Schmidt (2014) insists that McGee's ideas on water resemble, ironically, the proto-hydro-social. McGee acknowledged water's agency and developed his own concept of a hydro-social cycle, whereby 'the full control of water is actually a requirement for co-evolutionary progress' (Schmidt, 2014: 230). This indicates an anthropocentric point of view that contrasted the ideas of human–nature dualism given at that time (Schmidt, 2014: 230). As such, McGee cannot be associated as a co-founder of modern water but as someone who recognised people's dependency on and rights to it (Schmidt, 2014: 228). This example allows us to scrutinise the historicity of the hydro-social cycle and even more so to ask how the assessment of a 'water concept' does not succumb to a purely interpretative task—by chance failing as it did in McGee's case.

A second concern complains about the hydro-social cycle as a misleading construct in itself. Because the theory over-emphasises humans, it reduces water to a *thing*, a material (H_2O) whose actions are either 'good' or 'bad'. Floods, for example, might be perceived as a disruptive force for humans, but for many non-humans floodwaters are accompanied with ecological benefits (Camargo & Cortesi, 2019: 3). Moreover, and again ironically, the cyclical 'harmonious' dialectical relations between water, humans and technology, combined with the ideas of 'flow', are unnecessary and even counterproductive. One of the main tasks of Linton's (2010)modern water, and the other founders of the hydro-social cycle, was to multiply understandings on water, not to reify the same logic they intended to challenge. Camargo and Cortesi (2019: 5) perceive these metaphors as being 'uncritically deployed, again and again, to explain water to a point where these problematic representations are reified. Insofar as such metaphors have lost their descriptive and narrative value, and acquired a heuristic one, they have led to a restrictive understanding of water'. Not all waters behave in a repetitive, predictable manner, and even less so in regards to anthropogenic climate change that either deprives areas of water or disrupts them with an unmanageable abundancy (Camargo & Cortesi, 2019: 5).

Apart from the inner conflict invoked by the merging of ANT and relational Eco-Marxism, the hydro-social cycle offers a mainly anthropocentric, capital and politically driven relation to water. This in turn opens up new ways of addressing an enquiry on water-related issues, although one must: what sorts of issues are of concern, or, put in

Latourian words, what is the matter of concern? The hydro-social cycle frames and actually answers this question a priori by stating that water flows towards money, and as such towards powerful actors, driven by the process of capitalisation. This sort of framing, as discussed above, does not leave much room for explanations other than 'mean and powerful' actors depriving 'many poor' of their waters. The hydro-social territories or networks as employed by Boelens et al. (2016) do indeed open up differences of power between and within each network and distribute this power. In addition, more than Marxian theories, hydro-social territories rely more on a Foucauldian understanding of power. At this point, I would like to offer a way to analyse water-related issues that builds on the ideas of Latour, the hydro-social cycle and hydro-social territories. This conglomerate of different theories and assumptions is meant to offer a systematic approach to empirical research, describing one possible way.

Much has been written that—unfortunately—may cause more disturbance than clarity. I cannot claim to suggest an easier theory, something that solves this metaphysical puzzle or could 'let go' either of ANT, relational Eco-Marxism or Foucault. Many of these ideas have to be explained, and their combinations scrutinised and criticised, in order to make the following easier. Boelens et al. (2016), in their article on the hydro-social territories, use the term *collective*, which is not explained and—again—seems to have been borrowed by Latour, or at least it should have been, if hybrids and networks are incorporated concepts. Anyhow, this ontology of a collective was a theme Latour (2004) elaborated on in 'Politics of Nature', whereby he tried to develop a social democratic collective of humans and non-humans living together under a social contract, similar to Rousseau. This collective marks the entry point where I would like to begin my enquiry.

Building Water Collectives A paradox one encounters with ANT and the hydro-social cycle lies in the fact that while the former disentangles itself from theories or frameworks, it still demands a set of strict rules. The hydro-social cycle, on the other hand, conducts its research on a set of presumptions but does not offer or suggest any 'tools', so to speak. This difference must have been made obvious by now, but unfortunately, this stark difference is not complementary; that is to say, one could use the hydro-social cycle as a theory and ANT as its method. Therefore, any conjunction of the two leads to misleading, frantic theories, similar to two puzzle pieces that simply cannot fit but have been cut to such a degree—just to fit—that they lose much of their original implications. Therefore, I disagree with the arguments presented above by Castree (2002) and Lave (2015). There is much more to Latour's work than has been taken into account by the hydro-social cycle. ANT is a difficult encounter for most human geographers, I assume, because one must embrace an 'anti-theory' in a field that is still mostly guided by Marxist theories. However, in my estimate, it nevertheless offers many useful lessons, albeit they are only *negative lessons* that can serve as a guideline for any researcher, even though one does not make an ANT study descriptive, myopic and free of theory. The hydro-social cycle, not convincingly, quoted Latour's (1993) 'We have never been modern' ad nauseam, missing out on a lifetime of his work. For example, in 'Politics of Nature' (Latour, 2004), the much criticised modern dichotomy of society and nature is met with an alternative ontology, namely one of collectives. In contrast to socio-nature, where humans and non-humans find themselves mushed together but still separated as distinct objects or subjects, a collective is a consciously, intentionally constructed network of humans and non-humans. Ideally, a collective—in the sense it is probably mostly associated with—marks a 'unit' guided by a hierarchy, which points at a certain goal, for example maximising profit or sustainability. In reality, interests diverge and the building of collectives happens unconscious of its probable negative outcomes. Take, for example, how technologies during the twentieth century and onward have proliferated to such an extent that they have harmed myriad non-humans and humans alike, whose worries and lives reside *outside* the interest of one collective. What is external to the collective or internal is decided in due process by human actors—at least in Latour's idealised version. The relevant human actors who exercise the political ecology of a collective are scientists, politicians, economists and moralists (administrators and diplomats will be put aside for this thesis), each of whom has a diverse set of tasks to fulfil in order to *stabilise* the collective. This can be guaranteed best by following a four-step procedure. The first is *perplexity*, which means gathering around a matter of concern (e.g., climate change), and secondly *consultation* (e.g., which actors are relevant in climate change?). These first steps are guided by the maxims 'How many are we?' or 'To take into account'. The point here is to make sure that as many actors (humans and non-humans) are enlisted as possible, and none shall be mistakenly forgotten or excluded from the collective who could pose a threat to the collective in the future. Because climate change has been ignored by industrial countries, aligned with the concern of scientists and non-Western actors neglected for too long, climate change and all its consequences have multiplied and intensified and now threaten the collective. For such a mistake never to happen again, and to disempower climate sceptics, for example, non-humans must be granted a place in the collective as one of its citizens—I will return to the notion of 'citizenry', because Latour did not wish to humanise non-humans but socialise them. The third and fourth steps in the due process of constructing a collective consist of *hierarchisation* and *institutionalisation*, both of which revolve around the question 'Can we live together?' After humans and non-humans have been gathered, they are now assessed if they are compatible with the entities already internal to the collective, or in other words, whether can they fit into the same hierarchy. After each entity has received its legitimate place in the common world—and others have been excluded—the new 'citizens'" shall be *institutionalised* to 'no longer question their legitimate presence at the heart of collective life' (Latour, 2004: 109). This iterative process, however, is never done and will always remain 'ill-formed," and what has been excluded for now may return in the future, as long as its inclusion does not destabilise the collective. In the case of water, Le Bourhis (2005: 428) translates the idea of composing water parliaments, which 'do, in fact, operate as sites of collective decision but, for the most part, their functioning still remains non-democratic. In this domain, as in that of the environment in general, the political debate is still taking shape.' Although Latour (2004, 2005) is known for giving a voice to the non-human, which is alerting for many, he insists on the socio-democratic need to also identify, recognise and incorporate as many relevant human actors. His main intention in 'Politics of Nature' was to designate the sciences' already inherent capability to 'speak' for non-humans, so that they could be represented democratically for the well-being of the whole collective. This, as Latour emphasises, was never meant to give identity to the non-human but to emphasise human's dependency on the non-human, which is becoming increasingly self-evident in the looming climate crisis. Still, this notion of 'granting speech' to the non-human should not ignore or neglect human actors.

A water collective, or water parliaments (Latour, 2004: 174), so to speak, endeavours to convoke the human and non-human alike. Returning to the actors involved in composing the collective, it is the scientists whose main task is to represent and "speak" for non-humans, who simply cannot represent themselves. Politicians, on the other hand, mainly have to decide who is a friend and who is a foe to the collective. Their job is to declare enemies. Economists calculate the value between goods and people, and moralists are the key reason why composing the collective always remains an iterative process. They must contend with answering whether the collective has more than one hierarchy—that may cause disturbance and conflict—and which entities should be integrated into the collective to serve the common end.

Latour's metaphysical exercise may only remain a socio-democratic effort that can never be realised. Nonetheless, I find this concept of a 'social contract' to be a useful template with which to work. We can set this Latourian idea of a collective as a blueprint to look for deviations, conflicts and discrepancies that mark the analysed collective unstable—and therefore unsustainable.

Researchers could engage in multiple ways with both envisioned and current collectives without committing to either the hydro-social cycle or ANT. For this case, Latour developed a didactic version of ANT: the cartography of controversy, which is a tool to explore, investigate and visualise contemporary socio-technical debates. For this purpose, Venturini & Munk (2022) explained in detail the process to which one must commit in order to display the multiple actor-based perspectives that revolve around a controversy. The main reason why I would recommend this method is because it encourages researchers in relation to the following three aspects. Latour demands 'just to observe' collective life, but hereby implies (Venturini, 2010: 260; emphasis added):

(1) You shall not restrain your observation to any *single theory* or *methodology*;
(2) You shall observe from as *many viewpoints* as possible;
(3) You shall *listen to actors'* voices more than to your own presumptions.

The openness of this approach might be overwhelming at first, but it allows for a more objective perspective on the matter at hand. As such, the hydro-social cycle, the water-system approach, the hydro-social territories and discourse analysis are all means to the same end: to multiply the angles of a study from which to observe, describe and illustrate controversy. None of the tools, methods, paradigms and theories can demand hegemony, none of them collides; they simply deploy their strength. The more there are, the better. The cartography of controversy certainly implies complicating a researcher's life, but the outcome is invaluable because it enables an eclectic approach combined with a holistic outcome.

2.6 Conclusions: The Many Perspectives on Water

What is water? Is it what we make of it, as Linton suggests, i.e., materially, cognitively, culturally and discursively? Or is water more than just H_2O, a chemical compound, a controllable resource, whose outcome is always measurable and predictable? The hydro-social cycle argues that water has agency, and agencies of different waters have always been characterised by a reciprocal process with humanity, which are recognised as relational-dialectical. In the last, roughly two centuries, this relationship has experienced a drastic turn according to the hydro-social cycle, in that water has become a thing, an object, which is managed according to techno-managerial dispositives. As water has been pushed exclusively to the natural sphere—not geographically speaking, of course, but ontologically—the relationship between it and humanity has become as antagonistic as is often portrayed in natural sciences. In a prey-predator scheme, humanity cannot help but interfere and disrupt the hydrological cycle. Modern water has in this sense also co-created modern people: static, Neo-Malthusian people, overconsuming and simply unable to adapt—almost inanimate beings that resemble machines rather than humans. They constitute an important group of actors, however, who in my estimate should not be overlooked; these are the scientists who also contribute to the discourse on water. We, as hydro-social scientists, also depend on the production of facts by the natural sciences when we wish to present the status quo of these water-related issues.

Against this backdrop, I have chosen a similar approach to the aforementioned study by Flaminio (2021), i.e., identify different ontologies by working with the hydro-social cycle, and focus on sociotechnical controversies. In order to identify, navigate through and illustrate these controversies, as well as the involved human and non-human actors, I will draw back on Venturini and Munk's (2022) controversy mapping, which is mainly an ANT (actor–network theory) approach. At the same time, I will apply Linton's hydrolectics, which tries to identify relevant actors and associate them with different waters, in order to understand what people *think* water is and what sort of implications their ideas on water discursively and materially might have (Linton, 2010: 242). My focus lies primarily on the scientific discourse on Lake Urmia. Nevertheless, as the hydro-social cycle suggests, it is also necessary to historicise water—and thus the human-water relationship at Lake Urmia. Unfortunately, the history of Lake Urmia, to the best of my knowledge, is quasi non-existent in the English literature. Therefore, I will analyse the historical water management of Iran in general, from ancient times to the Islamic Revolution, through the lens of the hydro-social cycle.

References

Allan, J. A. (2005). Water in the environment/socio-economic development discourse: Sustainability, changing management paradigms and policy responses in a global system. *Government and Opposition, 40*(2), 181–199.

Badiou, A. (2008, December 2007). Live Badiou—Interview with Alain Badiou. In O. Feltham (Ed.), *Alain Badiou—Live theory* (pp. 136–139). Continuum.

Bakker, K. (2002). From state to market? Water *mercantilización* in Spain. *Environment and Planning a: Economy and Space, 34*(5), 767–790.

Bakker, K. J. (2003). A political ecology of water privatization. *Studies in Political Economy, 70*(1), 35–58.

Bakker, K. (2012). Water: Political, biopolitical, material. *Social Studies of Science, 42*(4), 616–623.

Balan, S. (2010). M. Foucault's view on power relations. *Cogito. Multidisciplinary Research Journal, II*, 55–61.

Barnes, J., & Alatout, S. (2012). Water worlds: Introduction to the special issue of Social Studies of Science. *Social Studies of Science, 42*, 483–488.

Bear, C. (2016). Socio-nature. In D. Richardson, N. Castree, M. F. Goodchild, A. Kobayashi, W. Liu, & R. A. Marston (Eds.), *International encyclopedia of geography: People, the earth, environment and technology* (pp. 1–5). Wiley.

Benjaminsen, T. A., & Svarstad, H. (2019). Political ecology. In *Encyclopedia of ecology* (pp. 391–396). Elsevier.

Birkenholtz, T. (2016). Drinking water. In P. Jackson, W. E. L. Spiess, & F. Sultana (Eds.), *Eating, drinking: Surviving* (pp. 23–30). Springer International Publishing.

Bleninger, T. (2010). *Environmental Planning, prediction and management of brine discharges from desalination plants: Final Report*, Muscat, Sultanate of Oman.

Boelens, R. (2013). The shotgun marriage: Water security, cultural politics, and forced engagements between official and local rights frameworks. *Water Security: Principles, Perspectives and Practices*, 239–255.

Boelens, R., Hoogesteger, J., Swyngedouw, E., Vos, J., & Wester, P. (2016). Hydrosocial territories: A political ecology perspective. *Water International, 41*(1), 1–14.

Brazil, R. (2020). *The weirdness of water*. https://www.chemistryworld.com/features/the-weirdness-of-water/4011260.article. Accessed 15 May 2024.

Camargo, A., & Cortesi, L. (2019). Flooding water and society. *WIREs Water, 6*(5).

Castree, N. (1996). Birds, mice and geography: Marxisms and dialectics. *Transactions of the Institute of British Geographers, 21*(2), 342–362.

Castree, N. (2001). Socializing nature: Theory, practice, and politics. In N. Castree & B. Braun (Eds.), *Social Nature: Theory, Practice, and Politics* (pp. 1–21).

Castree, N. (2002). *False antitheses? Marxism, nature and actor-networks*. Blackwell Publishers.

Castree, N. (2015). Changing the anthropo(s)cene. *Dialogues in Human Geography, 5*(3), 301–316.

Castree, N., & Braun, B. (Eds.). (2001). *Social nature: Theory, practice, and politics*.

Chang, H. (2014). *Is water H2O?: Evidence, realism and pluralism*. Springer.

Chatterjee, I., & Ahmed, W. (2019). Dialectical materialism: Marx's method in human geography? *ACME: An International Journal for Critical Geographies, 18*(2), 364–393.

Claus, C. A., Osterhoudt, S., Baker, L., Cortesi, L., Hebdon, C., & Zhang, A. (2015). Disaster, degradation, dystopia. In R. L. Bryant (Ed.), *The international handbook of political ecology* (pp. 291–304). Edward Elgar Publishing.

Clement, F., Suhardiman, D., & Bharati, L. (2017). IWRM discourses, institutional holy grail and water justice in Nepal. *Water Alternatives, 10*(3), 870–877.

Cortesi, L. (2021). An ontology of water and land in North Bihar, India. *Journal of the Royal Anthropological Institute, 27*, 870–889.

Damania, R., Desbureaux, S., Rodella, A.-S., Russ, J., & Zaveri, E. (2020). *Quality unknown: The invisible water crisis*. World Bank Group.

Dawoud, M. A. (2012). Environmental impacts of seawater desalination: Arabian Gulf Case study. *International Journal of Environment and Sustainability, 1*(3).

Dooge, J. C. i. (1983). On the study of water. *Hydrological Sciences Journal, 28*(1), 23–48.

D'Souza, R. (2016). *Drowned and dammed: Colonial capitalism and flood control in Eastern India*. Oxford University Press.

Dube, O. P., Brondizio, E. S., & Solecki, W. (2020). Introductory article: Technology, innovations, and environmental sustainability in the Anthropocene. *Current Opinion in Environmental Sustainability, 45*, A1–A6.

European Commission. (2019). *Executive summary of the fitness check of the water framework directive, groundwater directive, environmental quality standards directive and floods directive*. https://data.consilium.europa.eu/doc/document/ST-15101-2019-ADD-1/en/pdf. Accessed 16 August 2025.

Fernandez, S. (2014). Bringing politics back into water planning scenarios in Europe. *Journal of Hydrology, 518*, 17–27.

Flaminio, S. (2021). Modern and nonmodern waters: Sociotechnical controversies, successful anti-dam movements and water ontologies modern and nonmodern waters: Sociotechnical controversies, successful anti-dam movements and water ontologies. *Water Alternatives, 14*(1), 204–227.

Flyvbjerg, B., Ahlers, R., Zwarteveen, M., & Bakker, K. (2017). *Large dam development: From Trojan Horse to Pandora's Box—Oxford Handbooks*. Oxford University Press.

Gabriel, N. (2014). Urban Political Ecology: Environmental Imaginary, Governance, and the Non-Human. *Geography Compass 8*(1), 38–48.

Gallo, P., Amann-Winkel, K., Angell, C. A., et al. (2016). Water: A tale of two liquids. *Chemical Reviews, 116*(13), 7463–7500.

Gebhardt, H. (2016). Das „Anthropozän" - zur Konjunktur eines Begriffs. 28–42 Seiten / Heidelberger Jahrbücher Online, Bd. 1 (2016): Stabilität im Wandel / Heidelberger Jahrbücher Online, Bd. 1 (2016): Stabilität im Wandel.

Gleick, P. H. (2000). A look at twenty-first century water resources development. *Water International, 25*(1), 127–138.

Gleick, P. H. (2003). Global freshwater resources: soft-path solutions for the 21st century. *Science (New York, N.Y.), 302*(5650), 1524–1528.

Gleick, P. H. (2018). Transitions to freshwater sustainability. *Proceedings of the National Academy of Sciences of the United States of America, 115*(36), 8863–8871.

Gober, P., & Wheater, H. S. (2015). Debates-perspectives on socio-hydrology: Modeling flood risk as a public policy problem. *Water Resources Research, 51*(6), 4782–4788.

Gonda, R. (2023). Der Grand-Ethiopian-Renaissance-Staudamm. Das Recht auf Entwicklung und der Streit um den Nil. *Geographische Rundschau, 75*(3), 44–49.

Graefe, O. (2013). What is water? The history of a modern abstraction. *Social & Cultural Geography, 14*(1), 125–127.

Graefe, O. (2016). Integrated water resources management as a new approach to water security. In P. Jackson, W. E. L. Spiess, & F. Sultana (Eds.), *Eating, drinking: Surviving* (pp. 43–49). Springer International Publishing.

GWP. (2000). *Integrated water resources management* (TAC Background Paper No. 4).

Hamlin, C. (2000). 'Waters' or 'water'? Master narratives in water history and their implications for contemporary water policy. *Water Policy, 2*, 313–325.

Harvey, D. (2010). *Justice, nature and the geography of difference*. Blackwell.

Hastrup, K. (2013). Water and the configuration of social worlds: An anthropological perspective. *Journal of Water Resource and Protection, 05*(04), 59–66.

Hastrup, K., & Hastrup, F. (Eds.). (2015). *Waterworlds: Anthropology in fluid environments*. Berghahn Books.

ICOLD. (2011). *Constitution Statutus*. https://www.icold-cigb.org/userfiles/files/CIGB/INSTITUTIONAL_FILES/Constitution2011.pdf. Accessed 3 September 2024.

ICOLD. (2023). *General synthesis*. https://www.icoldcigb.org/GB/world_register/general_synthesis.asp. Accessed 3 September 2024.

Jameson, F. (2003). Future city. *New Left Review, 21*, 65–79.

Jones, E., Qadir, M., van Vliet, M. T. H., Smakhtin, V., & Kang, S.-M. (2019). The state of desalination and brine production: A global outlook. *The Science of the Total Environment, 657*, 1343–1356.

Krause, F. & Strang, V. (2016). Thinking Relationships Through Water. *Society & Natural Resources 29* (6), 633–638.

Laituri, M. (2020). The hydrosocial cycle in rapidly urbanizing watersheds. *Frontiers of Earth Science, 14*(2), 256–267.

Larsimont, R., & Martín, F. (2022). Chapter 4. Reflections from Latin America on the hydrosocial approach: Its use, abuse, and a possible way through the Maze. In C. Nicolas-Artero, S. Velut, G. Schneier-Madanes, F. Poupeau, & C. Chavarochette (Eds.), *Luttes pour l'eau dans les Amériques: Mésusages, arrangements et changements sociaux* (pp. 102–121). Éditions de l'IHEAL.

Latour, B. (1993). *We have never been modern*. Harvard University Press.

Latour, B. (1996). On actor-network theory. A few clarifications plus more than a few complications. *Soziale Welt* (47).

Latour, B. (2004). *Politics of nature: How to bring the sciences into democracy*. Harvard University Press.

Latour, B. (2005). *Reassembling the social: An introduction to actor-network-theory*. Oxford University Press.

Latour, B. (2013). *An inquiry into modes of existence: An anthropology of the moderns* (1st Harvard University Press paperback ed., 1st printing). Harvard University Press.

Latour, B. (2017). *Facing Gaia: Eight lectures on the new climatic regime* (C. Porter, Trans.). Polity.

Latour, B. (2018). *Down to Earth. Politics in the New Climatic Regime*. Polity Press, Cambridge.

Latour, B., & Callon, M. (2011). "Thou shall not calculate!" or How to symmetricalize gift and capital. *Athenea Digital, 11*(1).
Lave, R. (2015). Reassembling the structural: Political ecology and actor-network theory. In T. Perrault, J. McCarthy & G. Bridge (Eds.), *Handbook of political ecology* (pp. 213–223). Routledge.
Le Bourhis, J.-P. (2005). Water Parliaments some examples: Making things public, *Atmospheres of democracy.* MIT Press, 482–485.
Lefebvre, H. (1991). *The production of space* (33. Print). Blackwell Publishing.
Lemmens, P., & Hui, Y. (2017). Reframing the technosphere: Peter Sloterdijk and Bernard Stiegler's anthropotechnological diagnoses of the anthropocene. *Krisis, 2*, 26–41.
Lenton, T. M., & Latour, B. (2018). Gaia 2.0. *Science (New York, N.Y.), 361*(6407), 1066–1068.
Liao, Y.-K., & Schmidt, J. J. (2023). Hydrosocial geographies: Cycles, spaces and spheres of concern. *Progress in Environmental Geography.*
Linton, J. (2010). *What is water? The history of a modern abstraction.* UBC Press.
Linton, J. (2014). Modern water and its discontents: A history of hydrosocial renewal. *Wiley Interdisciplinary Reviews: Water, 1*(1), 111–120.
Linton, J., & Budds, J. (2014). The hydrosocial cycle: Defining and mobilizing a relational-dialectical approach to water. *Geoforum, 57*, 170–180.
Linton, J., & Krueger, T. (2020). *The ontological fallacy of the water framework directive: Implications and alternatives.*
Loftus, A. (2011). Thinking relationally about water: Review based on Linton's What is water? *The Geographical Journal, 177*(2), 186–188.
Marcinek, J. (2007). Hydrological cycle and water balance—A global survey. In J. L. Lozán, H. Grassl, P. Hupfer, L. Menzel, & C. D. Schönwiese (Eds.), *Global change: Enough water for all?* (pp. 33–37). Wissenschaftliche Auswertungen.
Mauss, M. (1990). *The gift: The form and reason for exchange in archaic societies* (Repr.). Routledge.
McGregor, D. (2012). Traditional knowledge: Considerations for protecting water in Ontario. *International Indigenous Policy Journal, 3*(3).
Meehan, K. M. (2014). Tool-power: Water infrastructure as wellsprings of state power. *Geoforum, 57*, 215–224.
Mehta, L., Alba, R., Bolding, A., et al. (2014). The politics of IWRM in Southern Africa. *International Journal of Water Resources Development, 30*(3), 528–542.
Mekonnen, M. M., & Hoekstra, A. Y. (2016). Four billion people facing severe water scarcity. *Science Advances, 2*(2).
Melo Zurita, M. d. L., Thomsen, D. C., Smith, T. F., Lyth, A., Preston, B. L., & Baum, S. (2015). Reframing water: Contesting H 2 O within the European Union. *Geoforum, 65*, 170–178.
Menga, F., & Swyngedouw, E. (2018). Water, technology and the nation-state.

Millington, N. (2018). Producing water scarcity in São Paulo, Brazil: The 2014–2015 water crisis and the binding politics of infrastructure. *Political Geography, 65*, 26–34.

Molle, F., Mollinga, P. P., & Philippus, W. (2009). Hydraulic bureaucracies and the hydraulic mission: Flows of water, flows of power. *Water Alternatives, 2*(3), 328–349.

Montanari, A., Bahr, J., Blöschl, G., et al. (2015). Fifty years of water resources research legacy and perspectives for the science of hydrology. *Water Resources Research, 51*(9), 6797–6803.

Moore, J. W. (Ed.). (2016). *Anthropocene or capitalocene? Nature, history, and the crisis of capitalism*. PM Press.

Mosse, D. (2003). *The rule of water: Statecraft, ecology and collective action in South India*. Oxford Univ. Press, New Delhi.

Mosse, D. (2008). Epilogue: The cultural politics of water—A comparative perspective. *Journal of Southern African Studies, 34*(4), 939–948.

Needham, P. (2000) What Is Water? *Analysis 60 (265),* 13–21.

Obertreis, J., Moss, T., Mollinga, P. P., & Bichsel, C. (2016). Water, infrastructure and political rule: Introduction to the special issue. *Water Alternatives, 9*(2), 168–181.

Ollman, B. (2003). *Dance of the dialectic: Steps in Marx's method*. University of Illinois Press.

Orlove, B., & Caton, S. C. (2010). Water sustainability: Anthropological approaches and prospects. *Annual Review of Anthropology, 39*(1), 401–415.

Oxford Learner's Dictionaries. (2024). *Definition of water noun from the Oxford Advanced Learner's Dictionary*. https://www.oxfordlearnersdictionaries.com/definition/english/water_1. Accessed 26 January 2024.

Paerregaard, K. (2018). Power in/of/as water: Revisiting the hydrologic cycle in the Peruvian Andes. *Wiley Interdisciplinary Reviews: Water, 5*(2).

Paerregaard, K. (2019). Liquid accountability: Water as a common, public and private good in the Peruvian Andes. *Water Alternatives, 12*(2), 488–502.

Paleologos, E. K., Al Nahyan, M. T., & Farouk, S. (2018). Risks and threats of desalination in the Arabian Gulf. *IOP Conference Series: Earth and Environmental Science, 191*, 12008.

Perera, D., Williams, S., North, T., & Curry, R. A. (2021) *Ageing water storage infrastructure: An emerging global risk 11.* United Nations University (UNU).

Pfister, L., Savenije, H. H. G., & Fenicia, F. (2009). *Leonardo da Vinci's Water Theory: On the origin and fate of water*. IAHS Special Publication No. 9.

Rattu, P., & Vèron, R. (2016). *Towards a Foucauldian Urban Political Ecology of water: Rethinking the hydro-social cycle and scholars' critical engagement.*

Ray, J. (1692). *The Wisdom of God Manifested in the Works of the Creation*. 2nd ed. London. Samuel Smith.

Robbins, P. (2012). *Political ecology: A critical introduction* (2nd ed.). Wiley.

Rosbjerg, D., & Rodda, J. (2019). IAHS: A brief history of hydrology. *History of Geo- and Space Sciences, 10*(1), 109–118.
Sayers, S. (1991). Marxism and the dialectical method: A critique of G.A. Cohen. In P. Osborne, S. Sayers (Ed.), Socialism, feminism and philosophy: A radical philosophy reader (pp. 40–169). Routledge.
Schmidt, J. J. (2014). Historicising the hydrosocial cycle. *Water Alternatives, 7*(1), 220–234.
Seguin, E., & Lord, L.-O. (2023). Bruno Latour's science is politics by other means between politics and ontology. *Perspectives on Science, 31*(1), 9–39.
Serres, M. (1982). *The parasite*. Johns Hopkins Univ. Press.
Seung-soo, H. (2018). *Water and disasters*. https://www.munichre.com/en/insights/natural-disaster-and-climate-change/interview-han-seung-soo.item-bf839e268fa0a1f926452c1e1097410b.html. Accessed 13 October 2023.
Sivapalan, M. (2015). Debates-perspectives on socio-hydrology: Changing water systems and the "tyranny of small problems"-Socio-hydrology. *Water Resources Research, 51*(6), 4795–4805.
Steffen, W., Grinevald, J., Crutzen, P., & McNeill, J. (2011). The anthropocene: conceptual and historical perspectives. *Philosophical transactions. Series A, Mathematical, physical, and engineering sciences, 369*(1938), 842–867.
Strang, V. (2004). *The meaning of water*. Routledge.
Strang, V. (2015). Reflecting nature: Water beings in history and imagination. In F. Hastrup & K. Hastrup (Eds.), *Waterworlds: Anthropology in fluid environments* (pp. 247–278). Berghahn Books.
Stubblefield, C. (2018). Managing the planet: The anthropocene, good stewardship, and the empty promise of a solution to ecological crisis. *Societies, 8*(2), 38.
Sultana, F. (2007). Water, water everywhere, but not a drop to drink: Pani politics (water politics) in Rural Bangladesh. *International Feminist Journal of Politics, 9*(4), 494–502.
Swyngedouw, E. (1996). The city as a hybrid: On nature, society and cyborg urbanization. *Capitalism Nature Socialism, 7*(2), 65–80.
Swyngedouw, E. (1999). Modernity and hybridity: Nature, regeneracionismo, and the production of the Spanish waterscape, 1890–1930. *Annals of the Association of American Geographers, 89*(3), 443–465.
Swyngedouw, E. (2004). *Social power and the urbanization of water: Flows of power*. Oxford University Press.
Swyngedouw, E. (2006). Circulations and metabolisms: (Hybrid) natures and (Cyborg) cities. *Science as Culture, 15*(2), 105–121.
Swyngedouw, E. (2009). The political economy and political ecology of the hydro-social cycle. *Journal of Contemporary Water Research & Education, 142*, 56–60.

Swyngedouw, E. (2015). *Liquid power: Contested hydro-modernities in twentieth-century Spain*. The MIT Press.

Swyngedouw, E. (2019). The anthropo(Obs)cene (A. E. Collective, T. Jazeel, & A. Kent, et al., Eds.; pp. 253–258). Wiley.

Tuan, Y.-F. (1968). *The hydrologic cycle and the wisdom of god: A theme in geoteleology*. University of Toronto Press.

Tvedt, T. (series editor), Jakobsson, E., Coopey, R., & Oestigaard, T. (eds.). 2006. *A History of Water*. 3 vols. London: I.B. Tauris. Volume 1: Water Control and River Biographies; Volume 2: The Political Economy of Water; Volume 3: The World of Water.

Tvedt, T. (2021). *Water and society: Changing perceptions of societal and historical development* (2nd ed.). I.B. Tauris.

Tvedt, T., & Oestigaard, T. (2011). *A history of the ideas of water: Deconstructing nature and constructing society*. I.B. Tauris.

UN. (2015). *Water for a sustainable world: Includes data and indicators annex for water and energy*. UNESCO.

UNDP. (2020). *Human development report 2020.: The next frontier. Human development and the Anthropocene*, New York.

United Nations. (2020). *Water and climate change: The United Nations World Water Development Report 2020*. France.

Venturini, T. (2010). Diving in magma: How to explore controversies with actor-network theory. *Public Understanding of Science, 19*(3), 258–273.

Venturini, T., & Munk, A. K. (2022). *Controversy mapping: A field guide*. Polity Press, Cam-bridge, UK, Medford, MA, USA.

Verma, K. C., & Kushwaha, A. S. (2014). Demineralization of drinking water: Is it prudent? *Medical Journal, Armed Forces India, 70*(4), 377–379.

Wesselink, A., Kooy, M., & Warner, J. (2017). Socio-hydrology and hydrosocial analysis: Toward dialogues across disciplines. *Wiley Interdisciplinary Reviews: Water, 4*(2), Article e1196.

Wester, P., Rap, E. & Vargas-Velázquez, S. (2009). The hydraulic mission and the Mexican hydrocracy: Regulating and reforming the flows of water and power. *Water Alternatives, 2*(3), 395–415.

Wester, P. (2009). Capturing the waters: The hydraulic mission in the Lerma-Chapala Basin, Mexico (1876–1976). *Water History, 1*(1), 9–29.

Wiegleb, V., & Bruns, A. (2018). Hydro-social arrangements and paradigmatic change in water governance: An analysis of the sustainable development goals (SDGs). *Sustainability Science, 13*(4), 1155–1166.

Wilson, N. J., Harris, L. M., Nelson, J., & Shah, S. H. (2019). Re-theorizing politics in water governance. *Water, 11*(7), 1470.

Wittfogel, K. A. (1957). *Oriental despotism a comparative study of total power*. Yale University Press.

Workman, C. L. (2019). Ebbs and flows of authority: Decentralization, development and the hydrosocial cycle in Lesotho. *Water, 11*(2), 184.

Worster, D. (1985). *Rivers of empire: Water, aridity, and the growth of the American West* (1st ed.). Pantheon Books.

Xu, L., Gober, P., Wheater, H. S., & Kajikawa, Y. (2018). Reframing socio-hydrological research to include a social science perspective. *Journal of Hydrology, 563*, 76–83.

Yates, J. S., Harris, L. M., & Wilson, N. J. (2017). Multiple ontologies of water: Politics, conflict and implications for governance. *Environment and Planning d: Society and Space, 35*(5), 797–815.

CHAPTER 3

The Waters of Iran: From a Romanticised Past to a Present Mutation

The history of water and the Iranian people may remind us of a four-book series, in which several genre characteristics can be found. At first, there is the epic fantasy novel of canal-digging gods, dragon-slaying kings banishing the evil drought and a hybrid goddess, half-human, half-waterfall, who nurtures all the rivers and their kin; deserts come to life and gardens spring from once-dead soil. The second book depicts a flourishing civilisation whose new landlords introduce a singular God as they prosper and expand their territory. Then tragedy strikes. External chaotic forces conquer and put an end to paradise, where the once advanced society regresses into a dark age, never to return to its former glory. The book closes. A man enters the scene, thus heralding the third book. His appearance differs drastically from his ancestors, indicating that there is a new age to come, a revolution, to be specific. He unveils the curtains behind him and promises a new utopia by revoking God, referring to ancient roots and legends while embracing the holy spirit of modernity, thus turning his back on agriculture to industrialise the impoverished nation. The promise turns out to be a lie, or at least a deceptive game, played, once again, by external forces. Book four, like déjà vu, opens with a revolution, and thus a promise, an antithesis to modernity, successfully knocking allegedly marionette king off the throne. The slogan declares, now conversely, to revoke the gods, re-embrace God and reintroduce quasi-traditions. However, the holy ghost of modernity still clings to

R. Gonda, *Lake Urmia's Hydro-Social Dilemma*,
https://doi.org/10.1007/978-3-032-08941-0_3

these aspirations like a shadow. More dams than ever are constructed, with prosperity through the hydraulic mission aspired. Yet again, though, the mission fails, as welfare and the water bodies throughout the country disappear—the former garden to become a desert again.

Within this certainly abbreviated, lurid comparison to a book series, one aspect becomes self-evident: water has played—and still plays—a major role in the constitution of Iran; or, as another scientist argued, water in Iran is a 'pivotal element of [...] social identity, cultural dynamics and belief systems' (Labbaf-Khaneiki, 2020: 5). As such, a historical analysis of Iran must include human-water dynamics, as omitting it would tell only half a story. The hydro-social dynamics of water, politics, religion and technology must be considered in order to identify the driving water concept, i.e., water ontologies. The goal of this chapter is not to summarise Iran's general history, which has been accomplished in innumerable works (see, for example, Cambridge's seven volumes on the 'History of Iran'; Axworthy, 2008; Briant, 2017; Brosius, 2021), or to describe the nation's historical water management (Mahmoudian & Mahmoudian, 2012; Saatsaz, 2020; Saatsaz & Rezaie, 2021; Wittfogel, 1957), but to highlight the heterogeneous water concepts given in each period and to analyse the human-water relationship in Iran and its many transformations from past to present. Little is known about the water concepts of ancient and present Iran, and it is not clear what factors have had an overall influence. To date, almost nothing from the perspective of the hydro-social cycle research, apart from Labbaf-Khaneiki's 'Cultural Dynamics of Water in Iranian Civilization' (2020), has been conducted. Clearly, however, aspects of ancient and modern Iranian water concepts are mentioned dispersedly across numerous disciplines such as agriculture, anthropology, archeology, history, political analysis, religious studies, rural development and so forth. In order to understand the complex nature of this topic, and to filter the ideas of water successfully, Tvedt and Oestigaard (2011: 11) highlight the need to 'break out of conventional analytical frames'. In this sense, this chapter will rely on a variety of disciplines. The aim is to gain an insight and understanding of Iran's, and thus Lake Urmia's, current water crisis, although not all the dynasties of ancient Iran can be fully covered herein. However, in respect to the comparison to the four-book series, I will analyse four different periods: the ancient, pre-Islamic period; the beginning of the Islamic period and its downfall; a transitional period and, finally, the modern period since the twentieth century.[1]

3.1 Early Beginnings—Mesopotamia, Elam and Achaemenid

Centuries before the Persian civilisation came into existence on the Iranian plateau and beyond, the pulse of the region was dictated by the Mesopotamian cultures, one of the cradles of human civilisation. The Mesopotamians' expertise in water management, in terms of administration, infrastructure and culture, had a significant impact on the Persians. The history of Mesopotamia was largely shaped by its alluvial plain, the river regimes of Euphrates and Tigris. Water management played a central role in all of these periods (Richardson, 2012: 11), as water was 'ever present in all aspects of Mesopotamian life, including religion, politics, law, economy, international affairs, war etc.' (Tamburrino, 2010: 32). From the 11th to the 4th millennium BC, the area experienced a profound transformation from small subsistence farms to an 'Urban Revolution' due to a drier climate, rising sea levels and the constant movement and vigour of the riverine system (Hole, 1994: 123; Verderame, 2020).[2] Agriculture at first did not necessitate centralised water administration, especially in northern Mesopotamia, and thus irrigation management was primarily a local affair (Adams et al., 1974; Wilkinson et al., 2012: 157; Yazdanpanah et al., 2013: 1613).[3] In the aftermath of 'environmental shocks' (Hole, 1994: 140), the landscape progressively urbanised

[1] I used the periods suggested by Saatsaz (2020) and Saatsaz and Rezaie (2021) as a template to structure this history chapter.

[2] The early stages of these agrarian societies benefited from favourable climatic conditions. In contrast to today, a monsoonal system provided large amounts of precipitation from the Indian Ocean to Mesopotamia, which therefore turned water in an almost ubiquitous resource. As a result, high water tables and an abundance of water allowed small-scale farms to settle dispersedly across the region and even supported "dry" farming (Hole, 1994: 122).

[3] Wilkinson et al. (2012) consider these early stages of manipulation of different fresh water sources and the establishment of networks—mostly in the form of a "herringbone" pattern type—a human niche construction. The small-scale operation led by local residents resulted in the emergence of local irrigation practices, which set the foundation for centralised, imperial irrigation management in the following millennia (Wilkinson et al., 2012: 157).

(Rost, 2017: 8). Powerful city-states such as Ur, Larsa and Babylon introduced central water management strategies, albeit to various degrees, and they either reacted to environmental challenges and/or used water for political and ideological purposes (Rost, 2017: 16–17).[4] Numerous technological inventions characterised Mesopotamian irrigation management (e.g., canals, dykes, reservoirs (Tamburrino, 2010)).[5] Mesopotamian royalty established both temples to oversee irrigation projects and state-financed hydraulic constructions, which demanded large labour forces and partly centralised control (Rost, 2017: 9). At the same time, these large-scale constructions, starting especially in the 2nd millennium BC, legitimised royal claim to power and radiated safety and prosperity (Rost, 2017; Winter, 2007).[6] In the polytheistic religion of Mesopotamia, water symbolised purity, fertility and the power to create—but also destroy—life. The worshipped deities of water, such as Inanna in Sumer and Istar in Babylonia, 'represented the threshold between life and death [...] [and] in contrast to other kinds of more specialized deities, were connected with the complete circle of life' (Saadi-Nejad, 2021: 41). The genesis of the Persian Empires, and as such the history of Iran, however, is

[4] In fact, some dynasties took a "hands-off" approach to irrigation management (Rost, 2017: 16–17). During the traditional water management period, water was a common pool resource and therefore of common interest. Although Hammurabi's Law, for example, regulated water management, it never targeted anything beyond primary canals or large-scale projects, '[...] as most state interventions were designed to provide and secure water supply for irrigation and not necessarily to manage the systems beyond the main canals' (Rost, 2017: 17). Only four paragraphs out of 282 addressed water law. The aim of the laws was to ensure the continuous, loss-free and productive operation of the agricultural sector. Farmers were encouraged to maintain their water infrastructure (e.g. dams, canals, barriers) and to use an appropriate amount of water without damaging their neighbours' crops (Kornfeld, 2009: 1). Rather than state control of water, customary law organised irrigation management because farmers were already economically interdependent.

[5] Geyer and Monchambert (2015) suggest that the technological ingenuity to build complex water networks might have been established first in the Mari area (northwest of Iraq and west of Syria) in the 3rd millennium BC, which could be considered the "ancestor" of water networks in the Near East that followed in the 1st millennium BC. In the case of Mari and subsequent city-states and rural regions, canals served at least four functions: irrigation of cultivated land, supplying towns, evacuation and drainage and navigation.

[6] Rost (2017: 17) argues that in earlier periods (e.g., Old Babylonian State 2017–1595 BC), 'state-sponsored hydraulic projects' were mainly implemented for existential reasons, yet they progressively 'served ideological and political ends'.

mostly associated with the Empire of Elam (Saatsaz & Rezaie, 2021: 4). As much as it was culturally influenced by the Babylonians, Sumer and the Assyrians of Mesopotamia, so did the Elamite influence the Persians, who migrated to the Iranian Plateau during the second half of the 2nd millennium BC. In fact, the first Persian civilisation was a 'hybrid between the native Elamites and intrusive Iranians' (Saadi-Nejad, 2021: 9), and so for this reason, the Elamite are considered one of the ancestors of Iran. The Persians even expressed their affinity with the Elamite by adopting their traditions and their script (Brosius, 2021: 17). For roughly two thousand years (2600–640 BC), Elam was a major political force in the southwest of present-day Iran, or, more specifically, in the province of Khuzestan (Brosius, 2021). The Elamite were determined to separate themselves from their Mesopotamian neighbours, and in this context, water, technology and socio-cultural dynamics were co-produced in so-called *ziggurats*. Although these ziggurats were not an Elamite but probably a Mesopotamian invention, the holy temples served a variety of functions. Ziggurats were stepped pyramidal monuments, ranging from three to seven storeys high, and they were exclusively accessible to priests, who made sacrifices at a shrine located at the top of the structure (Niroumand, 2012: 396). In present-day Iran, the most prominent and best-preserved ziggurat is the historical World Heritage Site of Tchoga Zanbil, located in the southwest of the country and established in 1250 BC during the reign of the Elamite (Saatsaz & Rezaie, 2021: 18). Moreover, Tchoga Zanbil marks the oldest water supply system in Iran (Saatsatz, 2020: 1751). Water gave meaning and a sense of sacredness to the place, while these sacred places provided orientation, harmony and spatial structure (Afkhami, 2019: 9). Even nowadays, Iranian scientists proudly state that 'Iranian' scientists constructed the world's first water purification plant (Hassanvand et al., 2017: 12), even though Iran or Persia as a country was still more than 500 years away from being formed. The construction of Tchoga Zanbil by the Elamite king Untash Napirasha was a religious and political act with the intention of liberating the people of Elam from the gods of southern Mesopotamia in order to shift the religious centre of gravity from Mesopotamia to Elam (Afkhami, 2019: 11). The other gods of southern Mesopotamia lost their grip on the Elamite, just as King Untash Napirasha intended (Afkhami, 2019)—a religious-political shift aided by 'pure' water and provided by innovative technology. However, the expansion of the temple was discontinued after the king's death, and the significance of Tchoga Zanbil declined shortly thereafter. Nevertheless, its historical imprint remained. Saatsaz (2020: 1751) argues that due to the ziggurat's complexity and the Elamite relation to the Iranians, Tchoga Zanbil could be marked as one of the most significant milestones in Iranian water management history.

However, as mentioned above, the Elamite Empire had a cultural influence on the Persians, albeit the latter originated from another region. Of all the Indo-Iranian tribes in Iran, it was the Persians, who lived in the southwest of Iran in the modern province of Fars, that built and expanded an entire empire on the sacred waters of Anāhitā. In fact, an analysis of the water goddess Anāhitā, one of the most important deities in Persian history, helps to draw several conclusions on Persian origins, water ontology and development. The Avesta (religious Zoroastrian text) describes Anāhitā as a woman wearing jewellery and 'garments of beaver' (Saadi-Nejad, 2021: 69–70). While the former indicates the influence of Mesopotamian deities and their emphasis on prosperity, the latter shows that she, as did the Indo-Iranians, came from a cold-climate environment (Saadi-Nejad, 2021: 70). The Indo-Iranians were in fact part of an Indo-Aryan tribe of wagon-equipped horse people who lived in southern Russia, on the steppes between the Dnepr and Volga rivers. During the middle or late part of the 2nd millennium BC, the Indo-Iranians split up and migrated to Europe, Central Asia, northern India and Iran. The Indo-Iranians migrated to the Iranian Plateau, and to this day they constitute the ethnic majority of Iran's population (Ehlers, 1980: 133; Frye, 1983). Due to her Indo-Iranian roots, Anāhitā is actually an Indo-European goddess, who shares many characteristics with Celtic deities. For example, all watercourses spring from heavenly water, or in case of Anāhitā, *she is a river* flowing down from the Milky Way but is simultaneously able to shapeshift between human form and a waterfall. Anāhitā's anthropomorphism caused her devotees to '[see] her as closer to themselves' (Saadi-Nejad, 2021: 51), suggesting that people conceptualised water as a higher-order power *and* as *someone* with whom they could identify. In other words, water was an integral part of people's lives, not only from a physical point of view, but also from a cosmological one: it had a significant social character. This hybrid understanding of water in Iranian culture proves a fundamental point in Linton's hydrosocial cycle theory, in that the concept of water (mainly represented by Anāhitā) transformed alongside the material history of Iran. For example, the Indo-European water goddesses were said to have different powers or functions, such as healing and fertility (Saadi-Nejad, 2021: 40–41).

However, Anāhitā deviates from her related goddesses, as she was additionally the deity of welfare and warfare. These additional traits resulted from the influence of Mesopotamian—or rather Babylonian—deities, who inspired the Persian royalty (Boyce, 1982: 39). Anāhitā's status was so encompassing that her devotees included the three major social casts (priests, warriors and producers) (Saadi-Nejad, 2021: 2). The concept of water was clearly bound to its religious aspects, which determined how people treated it. As the goddess Anāhitā embodied abundant, pure water, sanctity and danger, then 'water represented the threshold between life and death', or, put differently, 'the complete circle of life', in which case rivers, lakes and streams were considered sacred places of worship (Saadi-Nejad, 2021: 41). Inevitably, the Iranians had a great reverence for water, especially rivers, into which they did not allow their own people or foreigners to spit, wash their hands, bathe or throw cadavers, waste or the blood of their human or animal sacrifices (Boyce, 1982: 154; Saadi-Nejad, 2021: 124–125). Throughout Iranian history, Anāhitā played a role of varying importance (Saadi-Nejad, 2021). Nonetheless, the first well-known Iranian dynasty built their empire on a vast, inventive and well-administered water infrastructure, including the water goddess as an encompassing entity. The starting point of Persian dominance in the Middle East began ca 550/549 BC. At that time, Cyrus II, or Cyrus the Great, entered the history books with his victory over the Median King (another Iranian tribe, in northwest and western parts of today's Iran), which heralded the Achaemenid Empire (Brosius, 2021: 18). Cyrus the Great conquered numerous countries and ruled from Egypt and Anatolia in the west to the Indus Valley in the east (Axworthy, 2008: 16). Under the Achaemenid, the scattered Persian tribes transformed into a powerful, unified civilisation characterised by the same ideology of nation-building, and among many other aspects, they emphasised irrigated agriculture alongside the expansion of transport routes and urban areas (Ehlers, 1980: 135).

One of the greatest accomplishments during the Achaemenid era was the creation of a water bureaucracy led by the 'Water Master', or *Ao-Tar* (Mahmoudian & Mahmoudian, 2012: 94), while the king himself supervised major hydraulic constructions (Wittfogel, 1957) and was regarded as the 'Feeder' of his people (Darling, 2013: 36). The Avesta even promoted the expansion of cultivated land and the 'reclamation of wastelands' (Darling, 2013: 36). In this sense, the Achaemenid developed and constructed a long list of different technologies, such as ancient old gravity dams, irrigation canals, water reservoirs, diversion dams and watermills (Saatsaz, 2020: 1758f). Despite the fact that the Iranians settled in a favourable region—especially the Zagros Mountain area—most of the Iranian Plateau is characterised by an arid climate, which mostly favours a nomadic lifestyle. However, what set the irrigation management of the Achaemenid apart from that of the Mesopotamians and Elamites was the use of groundwater to cultivate and transform the hostile deserts into a habitable space. The technology to fulfil the Achaemenid ambition were the so-called *falaj*, *karez* or, more prominently, *qanats* (Yazdi & Labbaf-Khaneiki, 2013). Qanats are underground canals dug in mountainous regions to access groundwater aquifers. By tilting them slightly, and using gravity as a natural force, fresh water has been distributed across the Iranian plateau to provide drinking and irrigation water. Qanats have many advantages that are perfectly adapted to an arid climate, such as low evaporation loss, use of gravity instead of pumps, occasionally a higher discharge than modern concrete dams, the draining of saline and alkaline soils and aquifers, etc. (Ahmadi et al., 2010: 134). Their benefits were not only recognised by Middle Eastern people, but also spread to remote regions of today's China, Spain, Germany, Great Britain and the American continent (Balali et al., 2009: 97; Manuel et al., 2018; Mostafaeipour, 2010; Yazdi & Labbaf-Khaneiki, 2013: 45). Although scientists still dispute where and when exactly qanats were invented (Yazdi & Labbaf-Khaneiki, 2013: 46f), it is very likely that the Achaemenid were the first culture to successfully establish an entire empire on groundwater reservoirs—and thus qanats. However, next to the qanats, the Achaemenid Empire benefited from their variety of water technologies. Due to the sheer size of the realm, a range of climatic and social conditions challenged the hydraulic apparatus with regards to water issues. The Achaemenid water administration managed not only to solve these problems and keep revolts or mass migrations at bay, but also ensured the loyalty and admiration of many people. For example, Darius the Great prevented drought in the Indian part of Persia by initiating dam constructions. In Egypt, Darius ordered the digging of qanats to provide water for irrigation in the south and the construction of a canal to connect the Red Sea with the Mediterranean. In this way, he received the gratitude of the Egyptians and was henceforth entitled Pharaoh (Mahmoudian & Mahmoudian, 2012: 94). The Achaemenid Empire managed water to gain both power and—maybe its biggest achievement—form a national identity (Saatsaz & Rezaie, 2021:

8). While the empire was not established on major rivers (Saatsaz, 2020: 1776), it relied on qanats, which were organised primarily by local landlords (Yazdi & Labbaf-Khaneiki, 2013: 81). For the Achaemenid kings, as throughout the traditional phase, common people were regarded as capital. A larger population was associated with more agricultural production, higher tax revenues and thus higher state income (Labbaf-Khaneiki, 2020: 64). In this context, agriculture and (certainly) the efficient allocation and use of water were mandatory—an aspect that the Achaemenid mastered impressively because they '[...] could establish a strong link between science, technology, *and* culture' (Saatsaz & Rezaie, 2021: 7; emphasis added). The dependency on groundwater in the central Iranian Plateau prompted the Achaemenid government to subsidise, protect and maintain qanats (Yazdi & Labbaf-Khaneiki, 2013: 53). Protecting and maintaining the sensitive water infrastructure was similar to protecting the backbone and identity of the Achaemenid civilisation. Not surprisingly, Alexander the Great targeted and destroyed qanats in his conquest of the Middle East from 334 to 331 BC (Saatsaz, 2020: 1751), and during the Seleucid (312–250 BC) and Parthian (250 BC-150 AD), the formerly important qanat infrastructure was abandoned.[7] Both empires were not of Iranian origin, so this technology might have seemed foreign and unnecessary to them (Yazdi & Labbaf-Khaneiki, 2013: 55–56).

3.2 From Sassanid to the Arabs

The 'Persian revival' came as a result of the downfall of the last Parthian king, Artabanus IV, whose empire was under Hellenistic influence. Within a few years, Ardashir emerged as the new ruler of the newly founded Sassanid Empire. The king deliberately claimed to be of Achaemenid descent and therefore did everything in his power to restore the Persian identity (Axworthy, 2008: 45). In fact, the Sassanid originated from the same region (Fars) as the Achaemenid, whose 'elements [...] of governmental heritage survived through centuries of change' (Darling, 2013: 39). Similar to the Achaemenid, the Sassanians attempted to forge a distinct national identity that was defined, controlled, protected and spread by a central government. The central figure of the Sassanian social system was the God-King, symbolically coroneted by Anāhitā to

[7] Conversely, Ehlers (1980: 136) argues that even during the Parthians, qanat systems were further developed.

grant divine powers (Labbaf-Khaneiki, 2020: 3). Labbaf-Khaneiki (2020: 6) highlights that, according to Avesta, so-called 'Ashavan'[8] assured a constant flow of water through 'beautiful canals'. In the Sassanian rigid class-based society, the glorified monarch stood 'above all' but was responsible for maintaining the equilibrium by satisfying material and social needs as well as guarantee law and justice (Daniel, 2012). Among his duties was also his endeavour to provide prosperity by—among others—constructing dams and canals and extending cultivated land (Yarshater, 1983: 381). Although common people were subordinate to the aristocrats, the latter relied on *and* promoted population growth. In general, during the traditional phase of ancient Iran, the ruler's main interest was to encourage farmers to produce surplus value in order to generate tax revenues (Ehlers, 1980: 136) while keeping a relatively soft grip on farmers. In contrast to the Achaemenid ideology, however, the Sassanians emphasised 'control and sovereignty over their subjects' and the centrality of treasure, while prosperity for the rural areas or justice for the weak was non-existent (Darling, 2013: 41). Despite this oppressive state of mind, a king was considered unworthy if he caused unnecessary distress in his country (Darling, 2013: 41), hence 'water projects also demonstrat[ed] the state's concern for the well-being of its population' (Daryaee, 2023: 181).

During their ca. 400 years, the Sassanid tried to establish an urban dynasty. As such, they expanded urban areas and founded new cities across the empire (Axworthy, 2008: 49; Daniel, 2012: 60; Saatsaz & Rezaie, 2021: 8).[9] However, according to Frye (1983: 132) the expansion of urban settlements was less impressive than the expansion of cultivated lands, which laid the foundation for the urban dynasty. And although the Sassanid were experts in groundwater deployment and relied on qanats, among others, for their urban areas, the main emphasis was put on large rivers (Ehlers, 1980: 136). Despite the fact that Wittfogel's (1957)

[8] These Ashavan were probably associated with priests and kings to whom people owed gratitude for running water in springs and canals. 'Thus, Ashavanān's authority over water resources is deemed to be inherent in the divine creation of water' Labbaf-Khaneiki (2020: 6).

[9] Furthermore, the Sassanid rulers used resettlements as a political tool to shift a region's ethnic composition to their favour. In Mesopotamia, for example, canals were constructed from the Tigris to the east to supply the Persian colonial communities (Labbaf-Khaneiki, 2020: 62).

hypothesis of hydraulically based oriental despotism has often been criticised and has some empirical shortcomings, if it had to be designated to a period, it would be most appropriate with the Sassanids (Adams et al., 1974: 4). For the Sassanid, water management and political power were synonymous, while prosperity was based on a well-administered irrigation system and, more importantly, large-scale irrigation projects. Water was a central resource, whose control had to be established, for example through legal tests (Daryaee, 2023: 163). For this reason, the Sassanid emphasised a centralised water management system, which led to the instalment of the administrative water bureau, called *Diwan-e Kastfezoud* (Saatsaz & Rezaie, 2021: 8). The bureau recorded water rights in special documents, listed private or governmental farmlands for tax purposes, defined and executed official rulings on hydraulic infrastructure and maintained, repaired or dredged abandoned/destroyed qanats (Saatsaz & Rezaie, 2021: 8; Yazdi & Labbaf-Khaneiki, 2013: 56–57). Moreover, it also directed large-scale, comprehensively planned, state-controlled irrigation, for example in the empire's 'bread basket' Mesopotamia, where, at the time, the Sassanid relocated their capital to the city of Ctesiphon (south of Bagdad) (Axworthy, 2008: 35; Wilkinson et al., 2012: 169). Mesopotamia in particular exemplifies how a centrally led water bureau developed a highly engineered landscape, shaped by the idea of a grand, strict and rigid order—similar to the Sassanids' social ideology. To meet urban food demand and maximise production, the once variegated landscape in Mesopotamia was transformed into a continuous agricultural area. Former canals, which partly were 'indistinguishable [...] [from] natural alluvial streams' (Adams et al., 1974: 4), were surpassed by an unprecedented emphasis on straight, uniform, long, complex and branching secondary and tertiary canals. The uniformity of hydraulic structures in Mesopotamia confirms the centrally planned and administered character of the Sassanid's water bureau (Adams et al., 1974: 4). Other irrigation projects such as the ca. 300-km-long Nahrawan Canal, which enabled the irrigation of 8,000 km^2 of cultivated land, additionally exemplify the extensive administrative efforts of the centralised water bureau (Adams et al., 1974: 4f.; Rost, 2017: 6). Cultivated land expanded during the Sassanid period in an unprecedented manner, not to be matched ever since (Frye, 1983: 132), or perhaps only in very recent decades.

The encompassing involvement of the state in water affairs is nonetheless clearly evident. Wilkinson et al., (2012: 170), for instance, stress, that

'[…] on the topic of "water and power," […] concerning the relationship between system scale and imperial rule, […] imperial rule should not be assumed'. It is more likely that irrigation practices on the field, which expanded over ca 5,000 years, were integrated into the imperial system of the Sassanid.[10]

The presence of regional administrations across the empire facilitated the spread of innovation and technology. In fact, Greek (and especially Roman) technologies were incorporated into the Sassanian hydro-social cycle (Wilkinson & Rayne, 2010: 139). For example, captured Roman soldiers were forced to build dams, and Greek and Roman watermills were introduced and used to grind grain, which saved time, reduced costs, generated income, was sustainable and increased arable land. Until the mid-twentieth century, water mills played a crucial role in the socio-economic development of Iran (Saatsaz & Rezaie, 2021: 9f). Even though the Sassanians established a profound and complex water management system, and later even acknowledged the interdependence of social classes (especially warriors and peasants) (Darling, 2013), it may be the case that their dependence and reliance on agricultural income was the downfall of the empire. The more arable land was expanded, the more hydraulic infrastructure had to be built and, most importantly, maintained. The cost and labour required to maintain a constant flow of water was beyond the capacity of the water bureau, especially in times of war, when governments focused their revenues on the military (Adams et al., 1974: 5). As the river diversions in Mesopotamia entailed frantic work, a plague killed about one-third of the labour force, and 'war with Bzyantium increased the demand for revenue', following which food production declined (Darling, 2013: 43). Zarrinkub (1975: 17) adds to the causes of the Sassanids' collapse, namely a disruptive and dysfunctional gap between the ruling classes and the common people, greed, corruption and political interference by Zoroastrian priests, weakness and instability of the

[10] Despite the hierarchical structure in the feudalistic system of ancient Iran, or the top-down relationship between government and subjects (Yazdanpanah et al., 2013: 1613), which is highlighted by Wittfogel (1957), farmers organised themselves into so-called *buneh*—a multi-family network in which each peasant had the same social status and the goal to use water as efficiently as possible to maintain or increase the exploitation of cultivated land. In addition, *buneh* were interconnected within a broader network that included the landlord (or his representative), other *buneh* and specialists. The role of the state, on the other hand, was indirect and supportive (Balali et al., 2009: 97; Yazdanpanah et al., 2013: 1612).

government and warmongering, especially by the last kings, 'while treasury was exhausted and had become solely devoted to the adornments of vain pomp'.

The Islamic period after the fall of the Sassanid Empire is marked by a significant cultural transformation and the occurrence of diametric understandings of social systems, water management and water ontology. This transformation resulted in a specific Perso-Islamic culture. Daniel (2012: 66) highlights that the transition from Sassanian to Islamic Iran was both continuous and transformative, while subsequently 'Islam grew as a natural development of Iranian history in its regional setting'. However, as the following will show, this transformation, especially in relation to water, was dominated by Sassanian principles.

Shortly after the Arabs achieved the 'Victory of Victories' against the Sassanid in 642 AD (Zarrinkub, 1975), the new conquerors neither eradicated nor replaced[11] Iran's population but in pragmatic fashion substituted the ruling elites of the new territories (Gronke, 2016: 15). While the former Sassanid rulers maintained a rigid hierarchical structure, Islamic social dogma demanded a more humanistic and an egalitarian notion that emphasised 'justice, equality and fairness' (Saatsaz & Rezaie, 2021: 11), which allowed for vertical social movement (Ehlers, 1980: 154). However, the Abbasids, around 750, installed people from Khurasan—northeastern Iran—in their administration and army, 'rather than with their Arab compatriots who were infected with a *dangerous egalitarianism*' (Cahen, 1975: 308; emphasis added). Perhaps this shows that the Arab rulers were not interested in transforming Iran into a wholeheartedly Islamic, egalitarian country but adopted the Sassanid economic and administrative structure for plain 'fiscal reasons which are not difficult to understand' (Cahen, 1975: 313), since the old system was immensely profitable for the Arabs, albeit more onerous for the farmers. Darling (2013: 46) even states that the Arabs had become 'heirs of the Sasanians and their ideology', which highlights the persistence of Persian culture in Islamic Iran.

In this sense, the new Arabian rulers adapted the former water management system of the Sassanid, although it violated core principles of Islam and was—in theory—incommensurable with their own understandings of water. While water was abundant especially in the western and northern

[11] The Arabs never initiated the mass colonisation of Iran because the Arabian camel, the dromedary, was not suited for the cold climate of the Iranian plateau (Cahen, 1975).

provinces of the Sassanid Empire, the Arabs originated from the Arabian Peninsula, where it was comparatively scarce. They were, on the one hand, an urban and a religious society,[12] albeit, on the other hand, the Qu'ran did not emphasise agriculture (Ehlers, 1980: 154). Nevertheless, Islamic law extensively regulated water, which was mostly used for cultivation (Wilkinson, 1990: 63). People in the Arab Peninsula could only develop socio-economically, if they sustained, protected and therefore sustainably profited from the few natural resources at their disposal, such as water, animals and trees, surrounded by a harsh, arid environment. These differences between the Persians and Arabs become apparent by analysing the different religious perspectives on water. In Islam's cosmology, God created a world in which all natural elements symbolise the realm of the transcendental divine. Humans are part of an ecosystem that they should protect and perceive as a living, spiritual being. God entrusted the world to humanity, the starting point of their spiritual ascent (Al-Dhagistani, 2019). Öktem (2019: 127) argues that, at first, people were very affectionate towards and cared for the environment, as Muslims considered it a manifestation of God, full of gifts from the heavenly paradise, where even the rivers called 'Allah, Allah'. In Islam, water bears enormous symbolic power, since the *Shari'a*, Islam's religious norm, literally means 'path to the inexhaustible source of water' or 'access to the source of pure drinking water' (Naff, 2009: 40), which is synonymous with the path to spiritual enlightenment, salvation and to God himself (Ademi, 2019: 146–149). God provides rain only to those who remain on this path in reverence and worship ('Id as-Sahib, 2019: 151). Moreover, since water was a rather scarce resource in the Arabian Peninsula, the prophet Mohammed had very specific instructions to guarantee water access for all beings (including flora and fauna), maintain water quality and settle arguments.[13] Water was regarded as a common good, whereby 'no legal person or ruler [...] may appropriate a river, for example, or try and sell or rent its water' (Wilkinson, 1990: 60); to deprive others, especially those in need, would be disregarded by God, whereas only the benevolent, who

[12] Two aspects that were mutually reinforcing because Muslims could more easily follow their prescribed religious practices in cities (Ehlers, 1980: 144).

[13] The Prophet forbids the pollution of water, primarily of standing water, by e.g., urination or defecation. Sewage and water canals were to be strictly and distantly separated. Even from soldiers it was demanded neither to kill children, elders or women nor to pollute waters as well as overuse them during times of war ('Id as-Sahib, 2019: 154).

share their waters selflessly and provide water-scarce regions with water canals, will be close to God ('Id as-Sahib, 2019). Even the tax system for cultivated land, and hence the irrigation system, was relatively low, as Wilkinson (1990: 59) ironically comments: 'if the contribution can be called tax'. Since water preservation was mandatory, the Prophet ruled that irrigation had to be limited to 'an ankle's depth' (Wilkinson, 1990: 61). The Islamic water concept developed in an arid environment reflects the aforementioned ethics of justice, egalitarianism and fairness, coupled with religious endeavours.

In comparison, the aforementioned Achaemenid and Sassanid water concepts and practices were drastically opposed. Water and land were private goods, mostly possessed by the government, where especially the nobility profited from the extensive and taxed agricultural system. Take, for example, how during the Achaemenid period, a king privately owned an entire river and used its water as a means to present himself as a benevolent and grateful ruler, yet prosperous and full of splendour (Briant, 2017: 286; 301). Even the water goddess Anāhitā contrasts the Islamic water ontology: she, the waterfall—not just a transcendental 'thing' created by 'God'—runs both in summer and in winter, flowing down from heaven to earth into the Vourukasha Sea in great quantity, as from thereon she, the river, has 'thousand of bays and outlets' and 'a thousand men in depth' (Saadi-Nejad, 2021: 67). In exchange for sacrifices made to her, she offered power, especially of a military nature to those who fought the Lie.[14] The king, in this context, was the heroic central figure of Persian society. Darling (2013: 35) summarises the 'not quite divine, yet more than human' Achaemenian king's function as such: 'One purpose for having an all-powerful ruler was that subordinates would "feel fear" and so would remain each in his proper place'—a slogan the Sassanid kings surely adapted and enhanced, since they presented themselves as truly divine. Given these differences between the Islamic and Persian water concepts, it may be understandable why the new Arab rulers in Iran bent their land and water laws in the newly conquered country, whose private ownership and extensive irrigation system were not covered by Shari'a rules (Wilkinson, 1990: 58). Enforcing their Islamic rules, for example the 'ankle's depth' for irrigation, literally would have been, so

[14] The Lie was a Zoroastrian construct of the evil in the world that must be fought with the light, with the good led by Ahura Mazda and his earthly representatives, priests and kings.

Wilkinson (1990: 65) suggests, 'ruinous for the Muslims'. Therefore, since the Sassanid Empire was conquered entirely, and the transition was rather peaceful, the Arabs maintained the former Persian administration (Gronke, 2016: 15). In unison with the golden age, in the Islamic world, Iran experienced positive land and water developments from the seventh to the tenth century (Hashemi, 2012: 266–268). However, others note that, overall, the agricultural sector stagnated (Ehlers, 1980: 154), perhaps indicating the regression water management that followed in the next centuries, as suggested by Saatsaz and Rezaie (2021: 12):

> Despite all efforts made during this period, the lack of creativity and investment in promoting water-related infrastructure and technologies, occurring wars and territorial conflicts, prioritising economic and political concerns over social benefits along with poor water governance have resulted in water insecurity over centuries.

One of the reasons for the initial decline of agricultural systems was the new land tenure system introduced by the Arabs. Cities were the political and religious centres of Islamic Iran—and had been since ancient times. As such, political and economic power revolved around these urban centres, while the rural areas, and thus the agricultural sector, were subordinated to them; it was basically reduced to a supply system for the cities. The central government granted land to aristocrats, or the upper class in general, for a certain period (therefore, not as hereditary land). In this case, the landlord received a share of the income produced by the farmers, leading Bobek (1974) to coin the term 'rentier capitalism'. This system often led landlords to neglect investing in maintaining and extending irrigation infrastructure, since this would have been unprofitable from their perspective. Conversely, in order to increase their share, their primary objective was to increase production, which often led to the over-exploitation of cultivated land (Ehlers, 1980: 152–153).

3.3 Hydraulic Collapse and the Qajar

The decline of water management in Iran correlates with the dominance of nomadic cultures enforced by the new Turkish and Mongol rulers from the eleventh to the sixteenth century, which caused peasant life, and thus irrigation, to decline significantly (Gronke, 2016: 40; Ehlers, 1980: 154). After the Arab Caliphates dissolved in Iran, nomadic tribes such as the Seljuqs (eleventh to twelfth century), Mongols (thirteenth century) and later Timurid (fourteenth century) invaded Iran. These nomadic, mainly Turk, tribes sustainably changed societal and power structures in the centuries that followed and usurped the Persians from their seats of power—both political and military (Gronke, 2016: 39). During the Middle Ages, these invaders, especially the Mongols, destroyed dozens of cities, massacred millions and forced innumerable people into slavery (Petrushevsky, 1968: 482ff). Moreover, they destroyed irrigation works, including qanats, and at first showed little interest in developing agriculture, which led to a significant and sustainable decline in the agro-economy (Petrushevsky, 1968: 483; Roemer, 1989: 67–68; Yazdi & Labbaf-Khaneiki, 2013: 60).[15] As cities and villages were either demolished or abandoned, nomadism proliferated and established itself as a serious military power that always threatened and challenged the central governments' authority. Nomads, led by khans, now extended the—roughly speaking—duality of the urban bourgeoisie and rural proletariat as a third party (Ehlers, 1980: 155). Previously, nomads had existed during the Arab dynasties, but they had a positive reciprocal impact on agriculture. Cahen (1975: 309) clarifies this point:

> The Arab invasion is sometimes represented as ipso facto a penetration of nomads and thus, ipso facto a negative in its results and harmful to agriculture. Such a view may perhaps be partially true of the later invasions of the Turks and, even more, of the Mongols, but it was not so with the Arabs.

By contrast, the overall relationship between farmers and Arab nomads was one of 'mutual enrichment than of disorder' (Cahen, 1975: 310). The Turks and Mongols, on the other hand, were more interested in increasing their military power. The central governments in this case had the perpetual issue of satisfying both their administration monetarily and securing the military support of the khans. From medieval times to the twentieth century, Iran's '[...] land system was basically a bureaucratic, not a feudal system' (Lambton, 1969: 52), which emphasises an

[15] Nevertheless, Yazdi and Labbaf-Khaneiki (2013: 60) suggest that during the Mongol periods, the Ilkhanate, i.e., Persian ministers, later invested and supported the agricultural system. However, this was due to the fact that the Persian upper class and administration proved to be useful to the new rulers, so the Persian culture persisted somewhat while coexisting with the increasing influence of the Turks (Gronke, 2016: 50–51).

economic relation between landlords and farmers that secured the government's and the khans' income. From the tenth century, the permanent tax revenues derived from agriculture were insufficient to pay the military (Lambton, 1969). Therefore, the Seljuqs introduced a land tenure system that, according to Ehlers (1980: 155–156) and others, had far-reaching and drastic results. Similar to the Arab land system, the so-called *iqta system* gave real estates to royal family members, military servants, former officials and the ruler and his heirs. All of these actors had the sole interest to profit from agriculture as much—and as rapidly—as possible. Even if cultivated land was completely exploited, new land was awarded to the landlord to guarantee a steady income. This new system simultaneously fostered decentralisation, as landowners accumulated power and wealth, and therefore any central government with the aspiration to rule and expand had to cope with these centrifugal forces (Issawi, 1971; Lambton, 1969). Against this backdrop, many migrated to mountain areas to flee from wars and tax collectors (Ehlers, 1980: 157), and so agriculture lost much of its productivity. Furthermore, state revenues from agriculture in the fourteenth century were only one fifth in comparison to the Seljuqs, who were strongly influenced by Persian culture (Issawi, 1971): 'The Mongol invasion showed its effects for centuries, and it is questionable whether the country, […] ever regained the degree of prosperity that distinguished Iranian agriculture from the 4th/10th to the 6th/twelfth century' (Fragner, 1986: 491). While the glorified Sassanid era is often regarded as peaceful, stable, innovative and led by a powerful central government, the period following the Mongol invasion gravitated towards chaos, disorder and decentralisation (Marjanizadeh et al., 2012: 9).

From this perspective, the period from the eleventh to the sixteenth century in Iran marks a significant shift from a relatively prosperous period, supported by agricultural production, to the drastic decline and regression of irrigated agriculture. It is noteworthy that, despite the initial destructive nature of the nomadic tribes—even the Mongols—there was a subsequent interest in developing agriculture. Furthermore, all of these dynasties, which were usually short-lived, converted to Islam, and Persian culture was sustained during this volatile period (Petrushevsky, 1968: 483; Roemer, 1989: 142). Nevertheless, this period had lasting impacts on Iran, and arguably on its water management, as nomadism replaced the earlier sedentary structures of farmers and villages' complex irrigation systems were destroyed.

After eight and a half centuries of foreign domination, beginning with the Arabian Caliphates in the seventh century, once again, the new Empire by Safavid composed a centrally governed nation in Iran (Gronke, 2016: 68). Their relatively prosperous rule, which emphasised agricultural development, ended in the 1720s (Ehlers, 1980: 170–172), followed by another socio-economic and political decline, since Iran at beginning of the nineteenth century suffered 'depopulation, poverty, and economic exhaustion, and [was] largely isolated from mainstream of world politics, trade, science, and culture' (Issawi, 1971: 13).

3.4 Qajar Crisis

The Qajar period, which lasted ca 130 years (1789–1925), was marked by a variety of internal and external processes, since it emphasised heterogeneous political and economic particularism, while Western nations had a growing interest in Iran. Qajar rule was characterised by rather loose control over territory (Axworthy, 2008: 185). At this stage, water governance and legal development in Iran entered a transitional period, one that bridged the traditional period to the modern period beginning in the twentieth century (Saatsaz, 2020: 1760).

The Qajar kingdom, especially at the beginning, was not characterised by uniformity under the lead of a central government that clearly described and followed a national agenda, although centralisation increased in the course of the second half of the nineteenth century (Lambton, 1991: 469). It seems that, from the beginning of the nineteenth century to its end, social discrepancies grew out of proportion. In fact, Iran disintegrated spatially, economically and politically into autonomous and self-centred interest led by khans, large landlords, merchants, the ulama (Islam's clergymen) and family members of the royal house, as well as (corrupt) government officials (Ehlers, 1980: 182–184; Gronke, 2016: 89–90). These interests were economic in nature. In this context, water served primarily to increase tax revenues through agricultural production, which still laid the economic foundation of the nation. Although this economic relationship between land, water, farmers and the government was not new in Iran by any means, during the Qajar period, a so-called 'hydraulic collapse' occurred. Labbaf-Khaneiki (2020: 52) defined this phenomenon as 'a situation when socio-economic and political structures fall into an irreversible disorder that is in the first place sparked by water resources crisis'. These hydraulic collapses occurred in a cyclical manner in the face of the central government's inability to maintain power, which allowed other actors to increase their influence and thus burdened farmers with additional taxes. As a consequence, cultivated land fell fallow in many regions in the country (Wilkinson, 1990: 57–58). The Qajar period did in fact share many of the characteristics of the hydraulic collapse, since the central government was constantly in debt and due to its politics kept prosperity unintentionally in check and, until the end of the nineteenth century, brought ruin to the agricultural population (Lambton, 1991: 505; Seyf, 2010: 300). While water served Achaemenid and Sassanid ideologically to foster a national identity, the Qajar period lacked any comparable ambition to do so. They neither established a centrally administered water bureau (Darling, 2013) nor did they—with very few exceptions—finance large-scale water infrastructure or provide subsidies or incentives for farmers to improve access to and the allocation of water (Issawi, 1971; Lambton, 1991: 476)—ignoring the few, yet ineffective exceptions, for example at the beginning of the Qajar period (Ghasemi & Gheydari, 2019: 49). Newly constructed hydraulic structures such as dams were in a desolate condition, and due to bad timing and planning, they were regularly breached. Even the old dams from Sassanid times that still existed were in an overall better state. Fundamentally, the state had no interest in financing hydraulic infrastructure despite its economically beneficial payback (Floor, 2003: 174–180; Issawi, 1971) and its ability to form a national identity.

The Qajar kingdom, as the Safavid beforehand, engaged in numerous wars, especially against the Russian Empire, from the seventeenth to the nineteenth century. The last two during the Qajar period, however, proved to be fatal, as Iran lost its Caucasian territories (Axworthy, 2008: 185), which had negative consequences on the water supply system, since this area had an abundant water supply (Saatsaz, 2020: 1762). After the loss of Mesopotamia to the Ottomans during the Safavid era, a second water-abundant region was now excluded from Iran's hydro-social cycle, and therefore less water was available to produce both food and valuable goods such as silk and cotton (e.g., in the province of Gulistan) (Issawi, 1971). In the face of the technological and military advancements of the European, colonial powers that superseded the Iranians, a call for reforms was made which, however, was barely implemented due to the lack of central power of the government and the Shah's conservative state of mind. This internal struggle between modernising, or rather Westernising, Iran and traditional, Islamic values characterised the Qajar period (Gronke, 2016: 88–89). The Qajars' objective was to protect themselves both from internal (e.g., nomadic tribes, despots) and external (e.g., the British Empire, Russia) forces. In this context, they had to take out loans and were soon in a great deal of debt, as they were unable to collect sufficient taxes from the tribal leaders and corrupt officials. Even several tax increases, the sale of governmental positions and crown land (*khalesi*) and the commercialisation of agriculture did not provide the much-needed income (Issawi, 1971; Lambton, 1991). Moreover, the government's increasing need for money disrupted the alleged Circle of Justice, described extensively by Darling (2013) for the entire Middle East. The basic idea of this concept is that the relationship between rulers and subordinates is mutual in nature; that is to say, farmers provide tax revenues and rulers watch over their subordinates, protect them, provide them with prosperity and govern justly. During the course of Iranian history, there were numerous exceptions to this ideal; nevertheless, ruling powers were meant to re-invest parts of their income back into the social system. With the Qajar, this cyclical movement of money became unilateral, causing significant socio-economic disparities, decadency and the despair of impoverishing common people (Darling, 2013). If we accept Labbaf-Khaneiki's (2020) assumption that water was an important part of the Circle of Justice, then water lost its social competence.

However, as Foran (1989: 5) notes, the literature on economic development during the Qajar period is controversial. On the one hand, historians suggest a general upward trend of the economy, due to the inclusion of the agricultural sector into global markets and its subsequent commercialisation. On the other hand, as mentioned above, other historians argue that the economy stagnated, emphasising the Qajars' increasing social disparities, for example, between the rural and urban population (Lambton, 1991: 505). It has to be kept in mind that there were significant regional differences in the Qajar kingdom, so some regions did indeed experience positive developments during the nineteenth century. If a city or region succeeded in creating a prosperous economy, it was due to local leaders and the substantial social cohesion of semi-autonomous communities (see e.g., Labbaf-Khaneiki, 2020). Lambton (1969; in Issawi, 1971: 54) writes: 'In times of crises these [communities] have taken over under local leaders and carried on the administration, the religious classes in particular playing an important part in handing down of the traditions of society'. Lambton (1991) and Floor (2003), among others, also describe numerous cities and regions that prospered and profited despite an overall worsening situation for the majority of the population (Foran, 1989: 56).

As a reaction to the new market situation and Western demand for Iranian goods, agriculture was commercialised to sell cash crops such as silk, opium, cotton, fruits and nuts, all of which promised high revenues (Foran, 1989: 56; Lambton, 1991). In the past, farmers would not have a vested interest in their—from the state-loaned—land, agriculture turned into a business for the rich (Lambton, 1969). In addition, because the Qajar increased taxes, local or regional landlords, respectively khans, increased taxes as well, which threatened the farmers' existence. Hence, this exploitation was economically unbearable, and farmers had to borrow from, for example, their landlord or leave their land, as they could not provide taxes or an acceptable livelihood for themselves. Tribal leaders, rich merchants, large landlords, money-lenders, village headmen, prosperous peasants and foreign investors profited from this situation, as they bought more land to produce more cash crops (Lambton, 1991). As such, social imbalance and corruption increased. Lambton (1969) further adds that over time the spatial distance between urban landowners and rural population resulted in a lack of understanding and respect—a tendency that was amplified during the Qajar. Therefore, peasantry life experienced a state of decay because subsistence farming suffered in face of the increasing commercialisation of the agricultural sector, Iran's dependency on world markets and its increasingly 'peripheral position in the world system' (Foran, 1989: 36).[16]

Two events that perhaps exemplify the lack of social investments by the Qajar central government were the Great Famine, from 1869 to 1872, and the cholera epidemics which emerged in the seventeenth and eighteenth century, respectively in 1821. The Great Famine cost hundreds of thousands—perhaps even millions—of people's lives. Although the number of deaths vary drastically depending on the sources (from 200,000 to six million), there is no doubt that '[w]hatever the actual number of deaths might have been, famine certainly had disastrous effects

[16] A contrasting view on this state of peasantry life is offered by Floor (2003). Socio-economically, new goods and the penetration of world markets at the end of the nineteenth century on the whole increased peasants' well-being and while in the past 'they had become indebted to ensure their subsistence now [enjoyed] a higher quality of life' (Floor, 2003: 646). However, the general tendency seems to be that peasantry life regressed, which—next to other factors—caused civil unrest and led to the end of the Qajar period.

on Iran's population and its economy' (Seyf, 2010: 296). Although natural causes (e.g., droughts) contributed to the famine, Seyf (2010: 294) declares the Qajar central government and the commercialisation of the agricultural sector as causa prima:

> The Great Famine, far from being the result of a seasonal shortage of rain, was the culmination of the longer-term problem of Iranian agriculture, notably the failure to invest in the expansion and maintenance of irrigation works combined with the switch from food into cash crop production.

Similarly devastating was the effect of the cholera epidemics, especially in urban areas, which was a result of a weak central administration, the neglect of urban sanitation, religious and cultural factors as well as the ill-constructed water supply system (Afkhami, 1998). In 1796, Tehran was announced as the capital city of the Qajar (Ehlers, 1980: 172). Hydrologically, however, this decision was accompanied by increasing difficulties. Tehran had very few water sources and had therefore relied in the past on qanats to divert streams and underground springs from the Elburz range north of the city (Afkhami, 1998: 213; Knoblich, 2018). The 'heydays' of the qanats during the Qajar (Yazdi & Labbaf-Khaneiki, 2013: 61) were the hydrologic foundation of many cities, and in particular the increasingly growing capital city of Tehran.[17] In combination with a bad or rather non-existent sanitary system, however, the cholera bacterium infected and killed tens of thousands of people, with 20,000 alone in Tehran after the cholera epidemics of 1846 and 1852 (Lambton, 1991: 520), in part due to the qanat system. Although qanats were usually underground, within the municipal boundaries of Tehran they flowed openly in channels, or at least very close to the ground—especially in poorer districts. Lack of proper sewage and waste disposal, open defecation into gutters, cemeteries close to waterways and body washings of the infected dead spread the disease (Afkhami 1998: 213). Against this backdrop, Islamic concepts of water's purity proved also to be fatal. Neither running waters nor standing waters in, for example, home water reservoirs with a volume of at least one korr (350 litres) were perceived as potential sources of infection, and conclusively 'Asiatic Cholera [...] proved to be the singular Achilles' Heel in Islamic conceptions of purity'

[17] Tehran was the second-most populated city in Iran in 1876 with 85,000 inhabitants (Ehlers, 1980: 176).

(Afkhami, 1998: 215). However, under normal conditions cholera, would not have thrived, as the stomach's acids usually destroy the bacteria, but in combination with regular famines in the nineteenth century, this immune reaction failed to operate (Afkhami, 1998: 215).

As a result of these circumstances, the Qajar kingdom could not sustain its legitimacy. In essence, Darling (2013: 179) summarised the demise of the Qajar at the end of the nineteenth and the beginning of the twentieth century:

> Iran [...] exhibited signs of royal injustice: neglect of irrigation systems produced desertification, the treasury went empty while men of state grew rich, and Iranian territory and resources were prey to British and Russian exploitation. Iranian society, already economically stressed by increasing tax demands and incorporation into the world capitalist economy, massively protests the sale of the country's resources to Europeans. (Darling, 2013: 179)

This situation, aligned with resulting civil unrest, laid the foundation for a transformation of Iran which until this day has a significant influence on agriculture and the management of water.

3.5 The Pahlavi Period, Modern-Day Iran and the Hydraulic Mission

The twentieth century in Iran's water management illustrates how water turned into a resource for socio-economic development, or 'modern water', to use Linton's (2010) term. In general, and in contrast to the Qajar period, Iran experienced not one but two significant transformations within one century: first, from a traditional Irano-Islamic society to a superficially Westernised and yet thoroughly autocratic society, and second, to an Irano-Islamic society. However, during this volatile socio-political period, water management almost constantly alluded its trajectory, as pointed towards the elusive utopia of the—still ongoing—hydraulic mission.

Iran at the beginning of the twentieth century was defined by the continuity of medieval systems that had been sustained in large parts by the Qajar rulers (Lambton, 1991: 504). The lack of reforms, the extractivism of Western countries and the stagnancy of the economy, including agriculture, paralysed Iran socio-economically. In 1926, 12 million people lived in the country, mostly in peasant villages, tribes and small towns, while large parts of the population were illiterate and uneducated (Axworthy, 2008: 226). However, from 1925 to 1979, a new era began that was short-lived but substantial. With the coronation of the former Minister of War, Reza Khan, then Reza Shah Pahlavi, in 1925, the constitutional monarchy of Iran experienced drastic changes. The Reza Shah wished to break with Islamic tradition and thus intentionally named and portrayed himself as the heir of the Sassanid, and therefore pre-Islamic Iran, dynasty (Gronke, 2016: 100). He thus co-opted a new and rising social group that similarly wished to break with Islamic traditions and the social system, which they perceived as the main reason for the nation's socio-economic stagnancy and lack of modernity (Bill, 1970: 24; 36). This new 'professional-bureaucratic intelligentsia' received its education at foreign, Western universities, hence contradicting religious-based education traditionally taught in Iran (Bill, 1970: 25). As such, Reza Shah tried to emulate Western societies by emphasising education, expanding and improving infrastructure and introducing a Western lifestyle (Axworthy, 2008: 228f). At the same time, he ruled with an iron fist and governed with a control paradigm. He used his costly, national, modernised army to regain and centralise power internally (Heshmati, 1982: 31). Every potential threat was to be eradicated, and therefore he often violently forced nomads—at that time constituting 25% of the Iranian population—to become either sedentary or leave the country (Gronke, 2016: 100; Hambly, 1991: 225f). In this sense, Reza Shah was an autocrat who, in this way, 'affirmed both his absolute mastery of all his subjects and his commitment to modernity' (Hambly, 1991: 226). To finance his goals and expensive army (40% of government expenditure in 1920s), oil revenues marked an important source of income, which became the economic pillar of the country (Axworthy, 2008: 227). Nevertheless, despite Reza Shah's aspiration to strengthen Iran internally and stop foreign exploitation, he could not put an end to the extractivism politics exercised by the British. The Iranian government received a meagre 16–20% share of oil revenues, while socio-economic disparities increased in the country (Axworthy, 2008: 231). Moreover, the modernisation strategy of the Shah may have emphasised Western science and technology, urbanisation as well as industrialisation, and yet this process drove a wedge between traditional, religion-based society, the Shah and the new 'western' social class (Amanat, 2017: 387). Shah's development plans related mostly to neglected traditional agriculture and earmarked the beginning of modern water management that fell in line with the centrality, top-down and science & technology theme practised in the 'modernizing' Iran (Saatsaz, 2020: 1763; Yazdanpanah et al., 2013: 1613).

Agriculture received little attention in the early years of Reza Shah's rule. Conversely, disparity between small farmers and landlords increased in light of new reforms introduced in 1928/29, as corrupt bureaucrats expropriated smallholders and transferred their land to landlords (Heshmati, 1982: 31). Water governance from 1909 to 1959 was codified, and at first, the government formalised traditional and customary laws and practised a hands-off approach whereby groundwater was still governed by local stakeholders (Nabavi, 2017: 45–46). However, from the 1930s onwards, water governance, including groundwater, was successively redistributed to centralised institutions such as the Independent Irrigation Department under the Ministry of Agriculture. What in the Civil Code was just to be 'tacitly assumed' to belong to the state (Lambton, 1969: 211) was now clearly institutionalised in the form of ministries and departments. After thousands of years, local entities had to pass control over groundwater to central government, while traditional water management and technologies (e.g., qanats) became increasingly obsolete, since temporary scholars stigmatised them as remnants of an archaic, religion-based society, unsuitable for the modernisation efforts of Reza Shah (Yazdi & Labbaf-Khaneiki, 2013: 62; Nabavi, 2017: 47–50). In this 'first modernity', farmers devolved into passive users, lay expertise was belittled or ignored and water was converted into a private commodity (Yazdanapanah et al., 2013: 1613–1614). Moreover, agriculture was 'marginalised during that era', namely the era of the hydraulic mission that focused its efforts on urban areas and industrialisation (Yazdanapanah et al., 2013: 1613).

Against this drop, Reza Shah's son and successor Mohammad Reza Shah (1941–1978) continued the modernisation efforts of the Pahlavi dynasty. He co-operated closely with the US government, who had an interest in both Iran's oil resources and its advantage as a fortress against the Soviet Union and its communist ideology. Therefore, Iran's political stability and, as such, the Shah's preservation of power were listed high on the political agenda of the US government (Summitt, 2004). The new Shah's goal was therefore to propel development, mainly for the industrial sector, while maintaining his power position (Theberge, 1973: 10). In this sense, he wished to merge two contradictory ideologies: that of a benevolent monarch and of a revolutionist, capable of convincing the leftist and nationalist movements in the country to take his side, without undermining his absolute power (Ansari, 2001: 3). Water management in this context played a pivotal role, since rapidly growing cities such as Tehran and Isfahan were in dire need of water for domestic and industrial use (e.g., the steel complex in Isfahan). The vast qanat system in Tehran that had been established throughout the previous centuries did not suffice to supply the increasingly consumerist population. Dams, canals and wells replaced the qanats and enabled urban dwellers in the wealthier parts of the city to use close to 500 litres of water per day per person, while in contrast, in rural areas, 50 litres per day per person were available in the early 1970s. Yet, water was still projected to be scarce for urban areas (Beaumont, 1974: 427–428). The influence of the US government furthermore accelerated a profound transformation of the agricultural sector that was progressively changed in order to introduce private, US-American agro-businesses to the country. Since the 1950s, plans had been announced that consisted, for example in the region of Khuzestan, of 14 dams, hydroelectricity and large-scale irrigation projects to cultivate ca one million hectares of land (Theberge, 1973: 12). Modernisation efforts progressed—but only superficially, and to the detriment of most Iranians.

Despite the almost socialist rhetoric of Mohammed Reza Shah, '[s]ocioeconomic modernization was not matched by and equivalent political modernization, [furthermore], [...] expectations of prosperity and justice formerly articulated in the Circle of Justice were now expressed largely in the political language of the West' (Darling, 2013: 190). The Iranians did not receive any political participation in a democratic sense, which they had demanded, but 'only' an industrial revolution (Summitt, 2004: 579). The Shah's politics had two major consequences: first, increasing inner political tensions with the *ulema* (clerics), who were anxious to keep Islamic traditions, and second, the neglect and demise of rural areas. While the former gave 1979s Islamic Revolution a distinct leading figure, namely Ruhollah Khomeini, the latter drove millions of desperate people into the hands of the ulema. This process was especially catalysed by the Shah's modernising White Revolution, beginning in 1963, which intended to pre-empt a violent, bottom-up revolution, as it occurred throughout Europe and Russia (Ansari, 2001: 2). Moreover, it was intended to cut with the archaic past of Iran, move the peasantry closer to the political elite and secure the Shah's power (Bill, 1970: 31).

While the White Revolution consisted overall of 12 programs, except for land reform, literacy corps and health corps, 'the other programs [did] little more than exist on paper' (Theberge, 1973: 11). From a water management perspective, the Land Reform, i.e., the core element of the White Revolution, had a significant impact that re-organised the agricultural sector—and thus rural areas. The propagated goal of the Land Reform was to dissolve the traditional feudal system, reduce landlords' influence to just one village and in the process reallocate land to every farmer (Summitt, 2004: 569). However, the opposite happened, in that 60% of the 30 million Iranians lived in rural areas and worked in the agricultural sector, but only 5% of the farmers became landowners (Beaumont, 1974: 426–427; Theberge, 1973: 10). The Land Reform was 'in fact preserving rather than uprooting the patterns of traditionalism' (Bill, 1970: 33), as the Shah sold the countryside to primarily US-American agro-businesses while somewhat maintaining feudalism to assure political support by the landlords (Theberge, 1973: 11). For the peasantry, the socio-economic consequences of the Land Reform were detrimental. Since most of the new smallholders received too little land to ensure economically viable cultivation, the majority abandoned their land, migrated to the cities and thus left the land fallow. Modern, regional agricultural co-operatives were meant to substitute traditional agricultural systems, but the conservative farmers, whose co-operation would have been necessary, mistrusted the new methods (Amanat, 2017: 580; Heshmati, 1982: 192). For the peasants, the disruption of the traditional feudal and inter-communal *buneh* system brought put them in an 'uncomfortable situation', in which they had no experience in terms of how to bargain with the new government agents that replaced the landlords (Bill, 1970: 34; Heshmati, 1982: 192). In contrast to the purely economic nature of the Western hydraulic mission, 'Islamic water law is largely customary, highly pragmatic, and *supple in its application of moral principles and guidelines*' (Naff, 2009: 44; emphasis added), which was lost with the dissolution of the *buneh* system and in the increasing involvement of the central government in water resources. Essentially, farmers exchanged one landlord for another, and as Wilkinson notes (1990: 58), 'what does it matter if a peasant owns his land if every input and output passes through government or merchant-controlled pricing systems?' The Land Reform, in summary, did more harm than good for most farmers. They lost their land, and thus their cultural identity, while overall, Iran lost its traditional socio-economic backbone and experienced 'cultural alienation' (Amanat, 2017: 582).

The pinnacle of governmental power over water was achieved by the enactment of the 'Nationalisation of Water Resources' in 1968, whose management was appointed to the Ministry of Water and Power (Ministry of Water & Power, 1968). Yet again, though, even the centralised, technocratic control paradigm of the Ministry, which perceived traditional irrigation systems as economically inefficient and an obstacle to modernisation efforts, did not improve the socio-economic situation in the rural areas, although multiple irrigation projects were implemented (Beaumont, 1974: 423f.; Nabavi, 2017: 50). Less than 1% of the rural population had access to piped water, while farmers received 'only 30 percent of national income' (Amanat, 2017: 580). The ineffectiveness of these projects, and implementing several—on paper—efficiency-driven water laws, is often attributed to the inattention given to social aspects that would have helped to implement many of the desired projects instead of relying on engineering solutions alone (Beaumont, 1974: 426–427; Nabavi, 2017: 53). Given these circumstances, rural exodus continued, agricultural production decreased and civil unrest surged (Amanat, 2017: 579; Theberge, 1973: 12). As a consequence, Iran was forced to increase its food imports drastically, up to 70%, at an estimated value of two billion US$ in 1977—in comparison, Iran spent 68 million US$ on food imports in 1970/71 (Heshmati, 1982: 196; Darling, 2013: 206).

At the same time, Iran's water resources have been under severe stress ever since. Although the government passed a law in 1966 on the preservation of groundwater, and surrounded itself with experts from the fields of law, hydrology and geosciences, groundwater exploitation by the farmers continued. The rationale for their action was that declining water tables left qanats completely dried out, and fines for drilling illegal wells in forbidden areas 'were not indexed to the [increasing] inflation rate', which rose in the face of Iran's oil exports and the devaluation of its currency—also known as the 'Resource Curse' or 'Dutch Disease'. Therefore, farmers took—and still take—a calculated risk by drilling tens of thousands of illegal wells (Madani, 2014: 317; 320; Nabavi, 2017: 50–53).

In retrospect, Mohammad Rezah Shah's White Revolution failed to deliver its promises, and Iran subsequently turned into a sell-out to western, especially US-American, businesses. As the government gained total control over all land and water resources, however, they sold land

and water to the private sector, which made government control rather elusive (Nabavi, 2017: 50; Wilkinson, 1990: 58). The cultural turn from traditional water management systems to modern, western-emulating methods kick-started the hydraulic mission even after the fall of the Pahlavi dynasty in 1979; a Shah, who in the final stages of his reign, was completely detached from the declining living standards, loss of cultural identity and the oppression of his despotic regime (Gronke, 2016: 104). In the end, the Shah's propaganda and policy undermined and destabilised his own ideology—and thus his political legitimacy (Ansari, 2001: 2; Darling, 2013: 191).

After the Islamic Revolution in 1979, Iran's hydraulic mission excelled. Westernisation by the Pahlavis was followed by a counter-development in the new Islamic Republic of Iran and its Supreme Leader, Khomeini. The resurgence of neglected rural areas and their people, the emphasis on food self-sufficiency and leftist (coupled with traditional Islamic) ideologies were the centrepieces of the new development policies (Nabavi, 2017: 54). Against this backdrop, new laws were enacted such as the 'Fair Water Distribution Law' in 1983, which sought to resolve uneven water availability in the country—intensified by privatisation processes under the Pahlavi regime—by legalising many of the previously illegally drilled wells (Nabavi, 2017: 57–58). The government's socialist rhetoric and policies, however, changed drastically after the Iraq War, which lasted from 1981 to1988 and involved a focus on military investment. As a result, the agricultural sector was once again undermined, rendering the objective of self-sufficiency unattainable. Consequently, Iran shifted its attention towards economic development, which led to the marginalisation of justice and leftist ideologies. Ironically, the Islamic Republic constructed significantly more dams after the Islamic revolution, and especially since the 1990s under the lead of President Rafsajani (1989–1997), who particularly emphasised the fast development of the agricultural sector (Nabavi, 2017: 58–59). Dam construction surged in the country (Nabavi, 2017: 66), making it the youngest 'large dam-nation' in the world ranking in 13th place. On average, the current 594 large dams are only 20 years old (median age is 19) and an average height of 41 m (Perera et al., 2021: 9). Dam megalomania, however, could not provide the targeted self-sufficiency, and on the contrary, it may have harmed Iran in the long run (Mesgaran & Azadi, 2018: 23–24). Since the late 1990s to early 2000s, however, Iran experienced the combined socio-ecological consequences of the hydraulic mission and severe droughts, indicating a 'looming water crisis' (Foltz, 2002; Madani, 2014). The Islamic Republic, moreover, proved to be a continuation of the Pahlavi dynasty despite its 'Islamic and republican ideology'. Top-down policies, ideological intolerance, an oil-based development paradigm and the prevailing middle class and technocrats are reminiscent of the former regime; for the majority of Iranian people, in other words, nothing changed from a socio-political perspective other than a Westernising-industrialising despot, who was exchanged with an Islamic-traditional despot emphasising industrialisation. Ultimately, the new Islamic Republic has failed to provide social justice in similar way to its condemned predecessors (Darling, 2013: 208–209).

3.6 Conclusions: Iran's Turbulent Hydro-Social History

Throughout Iran's history, water has always been associated with prosperity. From the early Mesopotamian era and the first Iranian settler, all the way through to today's Iran, water has played a pivotal role in providing economic stability. However, as this chapter has shown, this paradigm has had a rather volatile existence. While the Achaemenid and Sassanid dynasties, as well as the Arabic caliphates until the eleventh century, indeed emphasised irrigated agriculture governed by a centralised water administration, the decline of water management has persisted for almost a millennium. Although the influence of Persian advisors in the courts of the mainly Turkish rulers attempted to emulate the 'golden age' of water management, agriculture has lost much of its productivity, with the cyclical destruction of irrigation systems by ever-emerging foreign nomadic tribes contributing to this trend. Only since the Pahlavi dynasties, and especially since Mohammad Reza Shah, has the economic value of water regained its relative importance. This, however, may be the only resemblance to the heydays of ancient Iranian water management, because other than water's economic value, its traditional ideological and social aspects have been largely neglected during the hydraulic mission, even in the aftermath of—and perhaps even more so after—the Islamic Revolution. Perhaps one of the main differences between the modern period and the traditional one is that the latter was characterised by the coexistence of local water management systems organised by the *buneh* and centralised water bureaus, while the former—during the hydraulic mission—sought to dissolve traditional water management, substitute it with a foreign paradigm and eventually control it centrally. The modern period, even more so, saw the reduction of rural areas and the sale of cultivated land to foreign investors. Put simply, the modern paradigm has drawn a sharp line with the traditional modus operandi and introduced a foreign, quasi-modern system that has subsequently acquired a hegemonic status.

Moreover, water in Iran has been transformed from an ontologically rich element to something that one might characterise as *modern water*. Specifically, in traditional, pre-Islamic Iran, water was anthropomorphic—as exemplified by the water goddess Anāhitā, who is considered one of the most important deities that still persists in some traditions of modern Iran (Saadi-Nejad, 2021: 159ff). The first ontological split of water, so to speak, occurred with the introduction of the Islamic God, whereby water was transformed from a divine, sacred entity *in itself* to a *creation* of a divine being. Following this first ontological, and I would argue fundamental, split between water and the divine, Islamic ethics have contributed over time to the widening gap between Iranians' perceptions of water and its intrinsic value. It has turned from a sacred entity into some*thing* created by a more distant God, in this case Allah. Despite this miniscule difference, water was no more anthropomorphic but successively objectified. For centuries, the conceptualisation of the environment in Islamic countries has faced cultural and ontological alienation and has been successively reduced to a mere resource that supplies nothing but anthropogenic needs (Öktem, 2019: 129ff). This anthropocentric scheme has found better tools in 'modern' Western technologies, which similarly prioritise human water needs over nature's—or less generalised—non-human water needs. The now obsolete and complex water management system of the past had to give way to a type of water which has been mostly stripped of its social dynamics and history. The cost of this reductionism may partly explain why Iran, and thus Lake Urmia, is facing not only an isolated material crisis, but also an ontological crisis that is reciprocally related to the material emergency.

References

Adams, R. M., Downing, T. E., Farrington, I., et al. (1974). *Irrigation's impact on society*. University of Arizona Press.

Ademi, C. (2019). Scharia - Streben nach dem Wesen der Dinge. Wasser als Vorbild. In S. Binay & M. Khorchide (Eds.), *Islamische Umwelttheologie: Ethik, Norm und Praxis* (pp. 139–149). Verlag Herder GmbH.

Afkhami, A., A. (1998). Disease and Water Supply: The Case of Cholera in 19th Century Iran. *Yale School of Forestry and Environmental Studies Bulletin Series 103*, 205–220.

Afkhami, B. (2019). Phenomenology, Chogha Zanbil Ziggurat. *Journal of Ancient History and Archeology, 6*(1), 5–13.

Ahmadi, H., Nazari Samani, A., & Malekian, A. (2010). The Qanat: A living history in Iran. In G. Schneier-Madanes & M.-F. Courel (Eds.), *Water and sustainability in Arid Regions* (pp. 125–138). Springer.

Al-Dhagistani, R. (2019). Der Sufi ist wie die Erde..." - Ein Versuch über die Grundzüge einer spirituellen Ökologie im Islam. In S. Binay & M. Khorchide (Eds.), *Islamische Umwelttheologie: Ethik, Norm und Praxis* (pp. 86–103). Verlag Herder GmbH.

Amanat, A. (2017). *Iran: A modern history*. Yale University Press.

Ansari, A. M. (2001). The Myth of the white revolution: Mohammad Reza Shah, 'modernization' and the consolidation of power. *Middle Eastern Studies, 37*(3), 1–24.

Axworthy, M. (2008). *A history of Iran: Empire of the mind*. Basic Books.

Balali, M. R., Keulartz, J., & Korthals, M. (2009). Reflexive water management in Arid Regions: The case of Iran. *Environmental Values, 18*(1), 91–112.

Beaumont, P. (1974). Water resource development in Iran. *The Geographical Journal, 140*(3), 418–431.

Bill, J. A. (1970). Modernization and reform from above: The case of Iran. *The Journal of Politics, 32*(1), 19–40.

Bobek, H. (1974). Zum Konzept des Rentenkapitalismus. *Tijdschrift Voor Economische En Sociale Geografie, 65*(2), 73–78.

Boyce, M. (1982) A history of Zoroastrianism, Volume 2 - Under the Achaemenians. In *Handbuch der Orientalistik*. E.J. Brill.

Briant, P. (2017). *Kings, countries, peoples: Selected studies on the Achaemenid Empire*. Franz Steiner Verlag.

Brosius, M. (2021) *A history of ancient Persia*. Wiley Blackwell.

Cahen, C. (1975). Tribes, cities and social organization. In *Bailey, Gray et al. (Hg.) 1975—The Cambridge history of Iran* (pp. 305–329).

Daniel, E. L. (2012). *The history of Iran* (2nd ed.). ABC-CLIO.

Darling, L. T. (2013). *A history of social justice and political power in the middle east: The circle of justice from mesopotamia to globalization*. Routledge.

Daryaee, T. (2023). *Sasanian Persia: The rise and fall of an empire*. I. B. Tauris.

Ehlers, E. (1980). *Iran: Grundzüge einer geographischen Landeskunde*. Wissenschaftliche Buchgesellschaft Darmstadt, Darmstadt.

Floor, W. M. (2003). *Agriculture in Qajar Iran*. Mage Publishers.

Foltz, R. C. (2002). Iran's water crisis: Cultural, political, and ethical dimensions. *Journal of Agricultural and Environmental Ethics, 15*, 357–380.

Foran, J. (1989). The concept of dependent development as a key to the political economy of Qajar Iran (1800–1925). *Iranian Studies, 22*(2–3), 5–56.

Fragner, B. (1986). Social and internal economic affairs. In P. Jackson & L. Lockhart (Eds.), *The Cambridge history of Iran* (pp. 491–567). Cambridge University Press.

Frye, R. N. (1983). The political history of Iran under the Sasanians. In E. E. Yarshater (Ed.), *The Cambridge history of Iran* (pp. 116–180). Cambridge University Press.

Geyer, B. & Monchambert, J.-Y. (2015). Canals and water supply in the lower Euphrates valley. *Water History 7* (1), 11–37.

Ghasemi, S. S. M., & Gheydari, A. G. (2019). Water and irrigation system in Qajar period. *Specialty Journal of Humanities and Cultural Science, 4*(2), 43–49.

Gronke, M. (2016). *Geschichte Irans: Von der Islamisierung bis zur Gegenwart, 5* (aktualisierte). Verlag C.H.

Hambly, G. R. G. (1991). The Pahlavi autocracy: Reign of Reza Shah, 1921–41. From Nadir Shah to the Islamic RepublicIn P. Avery, G. R. G. Hambly, & C. Melville (Eds.), *The Cambridge history of Iran* (Vol. 7, pp. 213–243). Cambridge University Press.

Hashemi, M. (2012). *A social-technical assessment framework for integrated water resources management (IRWM) in Lake Urmia Basin, Iran*. Doctor of Philosophy Thesis. Newcastle University, Newcastle.

Hassanvand, A., Parviz, M., & Gharibzadeh, S. (2017). Comparison between ancient water purification in Iran and new methods. *Journal of Islamic and Iranian Traditional Medicine, 7*(4), 12–.

Heshmati, M. (1982). *Die "weisse Revolution" und deren Wirkung auf die sozio-ökonomische Entwicklung Persiens: Anatomie eines gescheiterten Modernisierungskonzepts*. Haag + Herschen.

Hole, F. (1994). Environmental instabilities and urban origins. In G. Stein (Ed.), *Chiefdoms and early states in the Near East: The organizational dynamics of complexity* (pp. 121–152). Prehistory Press.

'Id as-Sahib, M. (2019). Wasserschutz in der prophetischen Tradition. Grundlegende Texte der Sunna. In S. Binay & M. Khorchide (Eds.), *Islamische Umwelttheologie: Ethik, Norm und Praxis* (pp. 150–165). Verlag Herder GmbH.

Issawi, C. (Ed.). (1971). *The economic history of Iran: 1800–1914*. University of Chicago Press.

Knoblich, K. (2018). *Mensch und Wasser: Eine elementare Kulturgeschichte*. Eigenverlag, Linden.

Kornfeld, I. E. (2009). Mesopotamia: A history of water and law. In J. W. Dellapenna & J. Gupta (Eds.), *The evolution of the law and politics of water* (pp. 21–36). Springer.

Labbaf-Khaneiki, M. (2020). *Cultural dynamics of water in Iranian Civilization* (1st ed.). Springer International Publishing.

Lambton, A. K. S. (1969) *Landlord and peasant in Persia: A Study of Land Tenure and Land Revenue Administration*, Reprinted (with new pref. and additional bibliogr.). Oxford University Press.

Lambton, A. K. S. (1991). Land tenure and revenue administration in the nineteenth century. From Nadir Shah to the Islamic RepublicIn P. Avery, G. R. G. Hambly, & C. Melville (Eds.), *The Cambridge history of Iran* (Vol. 7, pp. 459–505). Cambridge University Press.

Linton, J. (2010). *What is water? The history of a modern abstraction*. UBC Press, Vancouver.

Madani, K. (2014). Water management in Iran: What is causing the looming crisis? *Journal of Environmental Studies and Sciences, 4*(4), 315–328.

Mahmoudian, S. A., & Mahmoudian, S. N. (2012). Water and ancient water supply technologies in ancient Iran. In A. N. Angelakis, L. W. Mays, D. Koutsoyiannis, & N. Mamassis (Eds.), *Evolution of water supply through the millennia*. IWA Publishing.

Manuel, M., Lightfoot, D., & Fattahi, M. (2018). The sustainability of ancient water control techniques in Iran: An overview. *Water History, 10*(1), 13–30.

Marjanizadeh, S., Qureshi, A. S., Turral, H., & Talebzadeh, P. (2012) *From mesopotamia to the third millenium: The historical trajectory of water development and use in the Karkarheh River Basin, Iran* (IWMI Working Paper 135).

Mesgaran, M. B., & Azadi, P. (2018). *A national adaptation plan for water scarcity in Iran*. Stanford.

Ministry of Water and Power. (1968). *Iran water law and the manner of water nationalization*.

Mostafaeipour, A. (2010). Historical background, productivity and technical issues of Qanats. *Water History, 2*(1), 61–80.

Nabavi, E. (2017). (Ground)water governance and legal development in Iran, 1906–2016. *Middle East Law and Governance, 9*(1), 43–70.

Naff, T. (2009). Islamic law and the politics of water. In J. W. Dellapenna & J. Gupta (Eds.), *The evolution of the law and politics of water*. Springer Netherlands.

Niroumand, H., Zain, M. F. M., & Jamil, M. (2012). The important role of Chogha Zanbil in earth architecture based on material, structural and architectural aspects. *Advanced Materials Research, 457–458*, 395–398.

Öktem, K. (2019). Physische und metaphysische Selbstdefinition der Umwelt im Koran und die Verantwortung des Menschen. In S. Binay & M. Khorchide (Eds.), *Islamische Umwelttheologie: Ethik, Norm und Praxis* (pp. 115–135). Verlag Herder GmbH.

Perera, D., Williams, S., North, T., & Curry, R. A. (2021). *Ageing water storage infrastructure: An emerging global risk 11.* United Nations University (UNU).
Petrushevsky, I., P. (1968). The socio-economic condition of Iran under the Il-Khans. In J. A. Boyle (Ed.), *The Cambridge history of Iran Volume 5: The Saljuq and Mongol periods* (pp. 483–537). Cambridge University Press.
Richardson, S. (2012). Early mesopotamia: The presumptive state*. *Past & Present, 215*(1), 3–49.
Roemer, H. R. (1989). *Persien auf dem Weg in die Neuzeit: Iranische Geschichte von 1350–1750.* Steiner-Verlag, Beirut, Stuttgart.
Rost, S. (2017). Water management in Mesopotamia from the sixth till the first millennium B.C. *Wiley Interdisciplinary Reviews: Water,* 4(5), e1230.
Saadi-Nejad, M. (2021). *Anahita: A history and reception of the Iranian water godess.* I.B. Tauris.
Saatsaz, M. (2020). A historical investigation on water resources management in Iran. *Environment, Development and Sustainability, 22*(3), 1749–1785.
Saatsaz, M., & Rezaie, A. (2021). *Water resources management, technology, and culture in ancient Iran.*
Seyf, A. (2010). Iran and the Great Famine, 1870–72. *Middle Eastern Studies, 46*(2), 289–306.
Summitt, A. R. (2004). For a White Revolution: John F. Kennedy and the Shah of Iran. *Middle East Journal, 58*(4), 560–575.
Tamburrino, E. (2010). Water technology in ancient Mesopotamia. In L. W. Mays (Ed.), *Ancient water technologies* (pp. 29–52). Springer Science+Business Media B.V; Springer.
Theberge, R. (1973). Iran: Ten years after the "White Revolution." *MERIP Reports, 18*, 3–22.
Tvedt, T., & Oestigaard, T. (2011) *A history of the ideas of water: Deconstructing nature and constructing society.* I.B. Tauris.
Verderame, L. (2020). The sea in Sumerian literature. *Water History 12*(1), 75–91.
Wilkinson, J. C. (1990). Muslim land and water law. *Journal of Islamic Studies, 1*(1), 54–72.
Wilkinson, T. J. & Rayne, L. (2010). Hydraulic landscapes and imperial power in the Near East. *Water History 2* (2), 115–144.
Wilkinson, T. J., Boucharlat, R., Ertsen, M. W., et al. (2012). From human niche construction to imperial power: Long-term trends in ancient Iranian water systems. *Water History, 4*(2), 155–176.
Winter, I. J. (2007). *Representing abundance: A visual dimension of the Agrarian State.* In:
Wittfogel, K. A. (1957). *Oriental despotism a comparative study of total power.* Yale University Press.

Yarshater, E. (1983). The Seleucid, Parthian and Sasanian Periods. *The Cambridge History of Iran, 3*(1).

Yazdanpanah, M., Hayati, D., Zamani, G. H., Karbalaee, F., & Hochrainer-Stigler, S. (2013). Water management from tradition to second modernity: An analysis of the water crisis in Iran. *Environment, Development and Sustainability, 15*(6), 1605–1621.

Yazdi, A. A. S., & Labbaf-Khaneiki, M. (2013). *Veins of desert: A review on the technique of Quanat/Falaj/Karez* (2nd ed.). Iran Water Resources Management Company.

Zarrinkub, A. A.-H. (1975). The Arab conquest of Iran and its aftermath. In Gray Bailey et al. (Hg.), *1975—The Cambridge history of Iran* (pp. 1–57).

CHAPTER 4

Rethinking Lake Urmia: From a Matter of Fact to a Matter of Concern

In Chapters 3 & 4, I made two grand moves: first, a paradigm shift on the global water crisis by introducing the hydro-social cycle, and second, I discussed roughly 8,000 years of Iran's history in just one chapter. From now on, this thesis will change gear by moving closer to the sciences and the *matter of concern*—and it is this expression I would like to concretise.

Following Latour (2005), Lake Urmia and scientists share a very particular relationship. The desiccation of the lake, as well as its new, mutated form, have turned the once turquoise solitaire of the Azerbaijani region into a collective threat—and scientists treat this new entity as a matter of concern. The term is defined by Latour (2005: 244) as follows:

> 'An expression invented to contrast with matters of fact and to recall that ecological crises have no bearing on a type of beings (for example, nature or ecosystems) but on the way all beings are manufactured: the unexpected consequences as well as the mode of production and the manufacturers remain tied to matters of fact, whereas they appear to be detached from objects properly speaking'.

In other words, Latour wishes to sensitise for the social process of knowledge production, which opposes a rather simplistic, detached and entirely positivistic perception of Science (Chapter 2.1.), where *matters of fact* are produced from the very outset—an unwavering

R. Gonda, *Lake Urmia's Hydro-Social Dilemma*,
https://doi.org/10.1007/978-3-032-08941-0_4

trajectory pointing to 'things as they are'. On the contrary, the production of knowledge is full of conflict, and thus controversies. Even in the sciences there are certain topics that are constituted by *second-degree objectivity*. Again, according to Latour, the term specifies the genealogy of knowledge production, in that scientists (sciences) gather around a new phenomenon, elucidate it from different angles and—literally—construct knowledge. However, the different viewpoints result in sometimes incommensurable objectivities, and therefore second-degree objectivity, which means that they all may be true (at least if they are *well*-constructed) but are yet unable to reach consensus among the majority of researchers. If they finally agree, then facts (*first-degree objectivity*) have been finally accomplished (Venturini, 2012: 798–799). 'Science', as Collins and Pinch (1993: 148) point out, 'works by producing agreement among experts', and this final stage of knowledge production must be institutionalised and remain unscrutinised; otherwise, no civic development is possible because—and I cannot stress this point enough—Latour has been mistakenly used as a leading figure of social constructivism in the face of the erupting science wars since the 1990s, which have turned him 'into a mere sociologist, adherent of a "social construction" according to which "everything is equal," objective science and magic, superstition and flying saucers' (Latour, 2013: 296). Objective (scientific) knowledge extracted from a chain of references is indeed attainable, but it takes time, expertise, investments, tools, debates, conflicts and, finally, controversies: 'The fact is that *no one* has the answers—this is why they have to be stabilized revised' (Latour, 2005: 138).

This is the main purpose of this chapter, i.e., to elucidate the controversies revolving around Lake Urmia's production of knowledge—the matter of concern rather than the by default undisputed matter of fact. What we—as human geographers or social scientists—know about non-human entities is, especially in the early stage of knowledge production, not primarily built on facts but on *faults*, to paraphrase the provocative title of an article by Venturini (2012).

To achieve this task, I illuminate the scientific terrain at Lake Urmia. How many articles have been published? Which scientific disciplines dominate the production of knowledge? Which topics are researched intensively, which less so? Which processes accompany the production of knowledge, and why might they matter? Which researchers and papers occupy central positions in the actor network? And, finally, which topics are considered controversial and likely to be constituted by different viewpoints, and hence differing *perspectives*, on water and nature? These questions will be tackled in the following.

Methodologically, I rely mainly on literature analysis (scientific papers, institutional reports, local and international news articles) and, to some extent, on my interviews conducted in the field at Lake Urmia to supplement certain viewpoints. The literature was approached, so to speak, with the STS-maxim, that is by 'delving into the "messy" nature/history [of the Lake Urmia case] rather than coherent after-the-fact narratives' (Bencherki, 2017: 8). The literature was read and will be presented on two levels, as suggested by Nimmo's (2011: 114) reading of ANT. The first level focuses on empirical data (what do we now know about the matter in question?), while the second isolates the agency of the text, the definition of subjects, objects, domains and the agency of the material (how is the case presented, and what does the text achieve in doing so?). Indeed, texts are not the ideal material for any ANT study because of their cleanliness; they mute multiple realities and heterogeneity and leave out all the complications that can arise in the construction of knowledge. Nevertheless, they are inscriptions, 'enactments of reality', that essentially attempt to bring some order to the chaotic common world (Nimmo, 2011: 114). Rather than seeing texts as simple objects whose only purpose is to connect the reader to distant entities, they offer a glimpse into the author's positionality, his or her worldview, among many other characteristics that in some cases are nothing but veiled by logic or objective rationalism (Nimmo, 2011: 114).

4.1 The Feasibility of Lake Urmia for Controversy Mapping

Initially, it is necessary to indicate in a simplified manner the type, and thus the feasibility, of the Lake Urmia controversy; in other words, to ask the plain question: is the inquiry a good controversy to study or not? To provide some structure to this assessment, Venturini and Munk (2022: 44) offer four distinct categories, represented as spectrums in Fig. 4.3. The spectrum ranges from unfeasible for controversy mapping to ideal, or at least feasible, for the task. The four categories to evaluate are: (i) the number of different viewpoints on the matter (from binary to multiple), (ii) relevance (from cold to hot), (iii) the proficiency of debates (from commonplace to specialised) and (iv) accessibility (from secret to accessible).

The localisation of the Lake Urmia controversy on the spectrum of each of these four categories may be to some degree subjective. Nevertheless, the main objective here is not to provide an exact assessment but a rough estimate of whether Lake Urmia reaches the minimum requirement to be classified as a 'good' or scientifically valid topic for controversy mapping.

While the first category, i.e., number of viewpoints, is difficult to grasp for now, it suffices to say that the controversies revolving around Lake Urmia find themselves in more than a simple, binary opposition. There are indeed multiple viewpoints, albeit they usually tend to lie somewhere between two extremes. As the following chapters will show, these viewpoints are in fact more nuanced and cannot be categorised in simple binary opposition. Therefore, I would argue that along the spectrum of viewpoints, Lake Urmia ranges somewhere in the third quarter (seen from left to right).

Easier to assess is the next category, relevance, since this can be traced and quantified by using, for example, digital data: trending hashtags on Facebook, Twitter and Instagram, published online news items or search engine trends. Since the focus of this inquiry falls on the scientific debate on Lake Urmia, I have reduced the public perception to a Google trend statistic. The graphs in Fig. 4.1 show two distinct lines: a volatile 'interest' line (green/blue) and a linear trend line (red). While the former indicates the web search interest on Lake Urmia relative to its highest popularity, which equals the value of 100, the latter proves that public attention has increased overall since 2004. In comparison to two other prominent salt lakes that have experienced a similar fate—the Great Salt Lake (USA) and the Aral Sea (Kazakhstan/Uzbekistan)—Lake Urmia seems to receive more interest from Google users. The other lakes, especially the Great Salt Lake, have declined in terms of Google users' attention. However, there is a marked difference between the three water bodies: despite the aforementioned declining interest, the Great Salt Lake and Aral Sea receive more attention, which is indicated by a relatively constant interest value between 20 and 40—ignoring the acute decline in case of the Great Salt Lake. Lake Urmia has reached a similar constant interest value only since 2014 and is still overall lower compared to the Aral Sea. This observation helps to conclude that public awareness on Lake Urmia may not have achieved the same international attention as the other two examples herein; nevertheless, public awareness is likely to increase, since the highest interest values are located in the second third of the time axis.

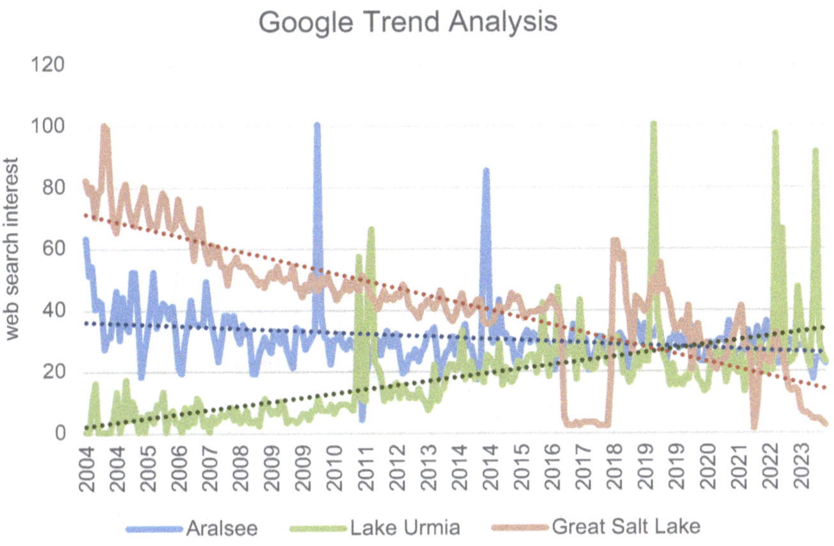

Fig. 4.1 Google search trends from January 2004 to November 2023 for Lake Urmia, Aral Sea and Great Salt Lake (accessed on 11th November 2023)

However, this indicator is of secondary, perhaps minor, interest, although it is assumable that Lake Urmia could be categorised as a 'hot' controversy. The primary indicator to confirm this tendency is by including the data on published scientific articles on Lake Urmia. This is done again in juxtaposition with the other two lakes, as illustrated in Fig. 4.2. Similar to the Google web search interest, more research has been conducted on the Aral Sea and Great Salt Lake—and increasingly so. In fact, the scientific community seems to have on recently taken academic interest in Lake Urmia since the early 2000s, which in turn means that knowledge on the lake has been produced only recently, while the other two lakes have been already researched more comprehensively. Because Lake Urmia has only lately turned into a matter of concern for scientists, the production of knowledge is relatively new and likely full of controversies, since there has not been enough time and resources to reach consensus on certain topics (the transformation from second-degree objectivity to first-degree objectivity). In total (as of November 12th, 2023), Aral Sea research has issued 2,225 articles, the Great Salt Lake 1,708 and Lake Urmia 1,008. However, publications on Lake

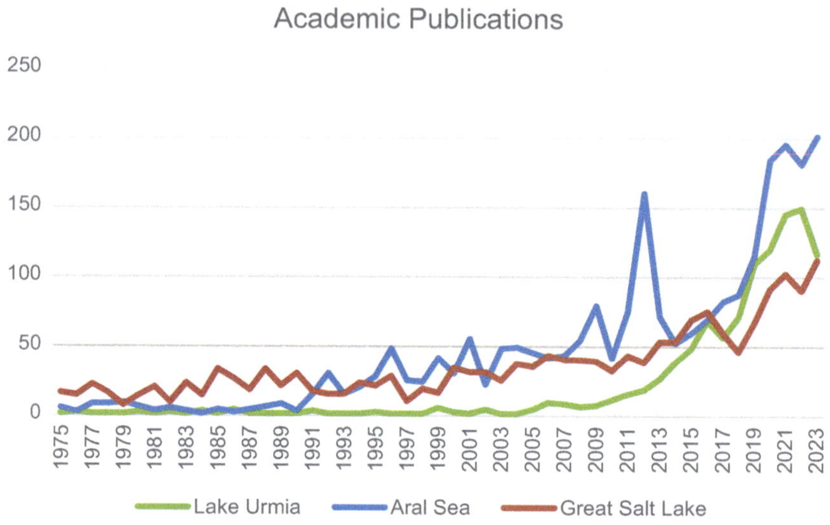

Fig. 4.2 Timeline of scientific publications per year from 1974–2022 for Lake Urmia, Aral Sea and Great Salt Lake; own illustration based on data from dimensions.ai (accessed on 11th November 2023)

Urmia have surged and passed the Great Salt Lake in annual publications since 2017, while they are approaching the same quantities as Aral Sea research. Given the information provided by Google trend statistics and publications data, I argue that Lake Urmia is undeniably a hot topic. Time will even tell whether the issue receives even more attention than it does right now—both public and academic. Therefore, I would put the relevancy not at the tail end of the relevance-spectrum but grant the phenomenon the possibility of reaching a higher status in public and academic awareness. In my opinion, Lake Urmia can be categorised as a controversy close to its pinnacle of relevance.

The third category, the proficiency of the controversy, can be answered briefly. Although Lake Urmia certainly marks a topic that is widely discussed by the Iranian public—online and in the media—and especially in the mostly affected provinces in the northwest of the country, the debates are also characterised by a high degree of academic involvement. In other words, as indicated above, there are plenty of scientific publications and experts on the matter. This way, the controversy mapping avoids badly constructed viewpoints that are established mostly on hearsay and thus lack expert knowledge in both theory and practice.

Lastly, I consider the accessibility of the controversy. This is a difficult topic in the context of Iran's tense political situation. On the one hand, research papers are written in the English language, and these are available to the public, albeit, alas, many are behind a paywall. However, direct access to local—or, in general, Iranian—experts on Lake Urmia is in most cases not possible, since in my experience they were not keen to participate in any form of interview (in situ during my field visits or remotely). Moreover, official documents are either impossible to access or offer only limited insights into the motives of the central government, its ministries or other state-run institutions. An additional limit was set by entry restrictions during the Covid pandemic, from 2020 to 2022, and the tense political situation in Iran, both of which made it difficult or perilous to conduct interviews with local experts, including lay experts such as farmers. Accessibility, therefore, is mostly reduced herein to a discourse analysis of scientific works. I have made efforts to include other, especially non-academic viewpoints such as the international and national media's perceptions of Lake Urmia, local farmers and relevant institutions. However, they function mostly as a minority report; that is to say, there are definitively gaps in knowledge or insights that I could not sufficiently fill. Nevertheless, since controversy mapping is considered by Venturini and Munk (2022: 50) as a '"secondary analysis" of the scientific debates', the focus of published papers on Lake Urmia shall (in my estimate) suffice for this thesis and its main objectives.

In summary, Lake Urmia can be categorised as a feasible controversy study that—despite some limitations—marks a fruitful and insightful matter of concern for the following and future inquiries (Fig. 4.3).

Fig. 4.3 Feasibility of Lake Urmia for controversy mapping

Given the nature of this inquiry, the next step will be to elucidate the scientific community that has gathered around Lake Urmia. As such, I will historicise scientific development in relation to the lake's desiccation. Furthermore, I will define the different periods through which research on Lake Urmia has gone, thereby providing information about the universities and, more importantly, the disciplines that participate in the discourse. However, it will become clear that none of the scientific developments can be seen in isolation, as many other actors are involved in the construction of knowledge about the lake.

4.2 Historicising Lake Urmia and Scientific Development

Figure 4.4 illustrates four different factors in the given time frame from 1992 to 2022: published scientific papers (represented as pillars), Lake Urmia's water level measured in metres above the Persian Gulf (the turquoise line), the four presidencies in Iran along the time axis and, lastly, events that will be elaborated in the following. As can be observed above, Lake Urmia has only been the subject of a growing number of studies for approximately two decades, at least according to the number of published papers. However, prior to 2013, the number of scientific insights on Lake Urmia was relatively limited. For example, only 16 papers were published between 2000 and 2013, in stark contrast to the 812 published between 2013 and 2022. This does not imply that scientists were previously unaware of Lake Urmia; rather, it suggests that as the crisis began to unfold, they began to focus their attention on the topic, publishing their articles and sharing their accumulated insights and data. As observed already by Parsinejad (2022: 4), scientific publications did not relate directly proportionally to the shrinking of Lake Urmia, or put simpler: the scientific interest in Lake Urmia came with some delay. From 1996 to 2002, water levels and.

the surface area decreased successively and significantly, i.e., within six years, the water level dropped by about 4.5 m, while the surface area shrank from 6,004.79 m^2 to 4,549.47 m^2. Interviewees at Lake Urmia, mainly farmers, stated that they had already noticed 15 to 30 years previously—so the early 1990s to early 2000s—how 'the shorelines were receding', to use their words, 'day by day'. Yet, scientific publications on the early stages of the phenomenon were rather scarce. Between 1996 and 2002, only eight papers were published on the issue, and only one of them noted that the surface of the lake had shrunk by 1,000 km^2

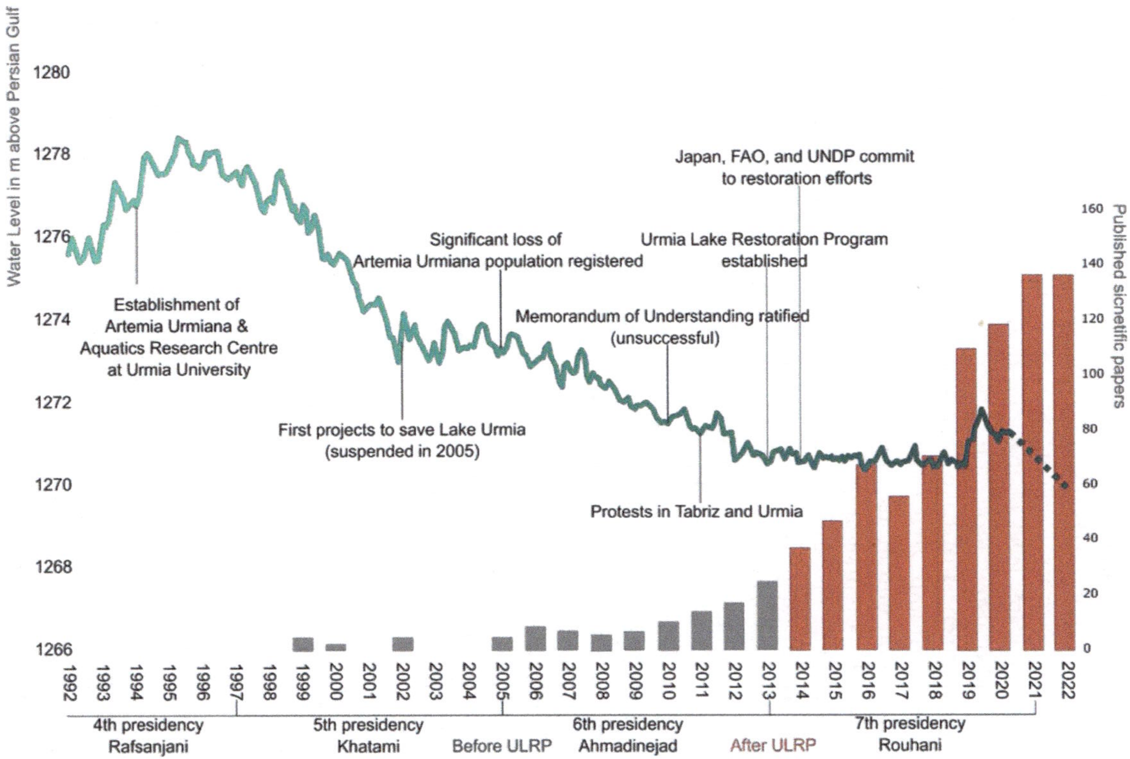

Fig. 4.4 Published scientific papers, Lake Urmia's water level, presidencies and relevant events

between 1998 and 2001. Aleshkeik and Ghorbanali (2004: Abstract) stated: 'It is necessary to say that the tide-range in Urmia Lake is inconsiderable. These changes derive from water balance of Urmia Lake'. However, nothing else was registered or further elaborated on the lake's water balance, whereas a specific aquatic entity indicated the upcoming anomaly of the water level's behaviour, or rather, the unfolding of a crisis and profound transmutation all together.

The distinctiveness of Lake Urmia is often associated with its equally unique brine shrimp *Artemia urmiana* (*A. urmiana*). By tolerating high salinity concentrations, the brine shrimp fills an ecological niche without which potential aquatic predators would be capable of surviving (Abbaspour & Nazaridoust, 2007: 166). Under normal conditions—that is, a certain range of saline concentration—*A. urmiana* population thrived in the past and served as a food source for migratory birds such as flamingos, which annually flew to Lake Urmia by the tens of thousands to breed (Sima et al., 2021: 8). Alas, despite the absence of local aquatic predators, *A. urmiana* s economic value on the global market [US$50 to 129 per kilogram (IRNA, 2022)] has led to a remarkably ironic pathway: it is sold as a fodder for fish, prawns and other marine animals (Azernews, 2019)—the very predators that would never have tasted *A. urmiana* if it were not for globalised markets. When the water level at Lake Urmia reached a historic high in the 1990s, the economic value of *A. urmiana* led to dreams and visions of an *A. urmiana* industry, 'in huge economical scale with intentions to enter the world market' (Golabian, 2011: 375). The relationship with this valuable resource was intensified even further when Naser Agh and Farzaneh Noor established the *A. urmiana* and Aquatics Research Centre at the University of Urmia in 1994, dedicated to the study of the local brine shrimp (Urmia University, n.d.). The high hopes placed in *A. urmiana* proved to be false after a decade, as the biotope, Lake Urmia, was likely to become uninhabitable due to another actor, namely salt, which is both a friend and a foe to *A. urmiana*, as it keeps aquatic predators literally at bay but in higher concentrations negatively restricts *A. urmiana*'s reproductivity (Agh et al., 2008). Therefore, when Lake Urmia began to dry, *A. urmiana* was one of the most important indicators of the lake's health. In an interview conducted in 2005, Agh noted a steady decline in the shrimp population through a study showing that instead of 250,000 cysts populating one cubic metre of water, fewer than 500 cysts were found. However, Agh admitted that there were no exact figures on the population of *A. urmiana* in Lake Urmia (Mehr News Agency, 2005). Nevertheless, the decline of the brine shrimp was one of the first troubling signs.

Two years later, in 2007, Eimanifar and Mohebbi (2007) published a brief review article when the lake did not seem to rebalance itself and continued its decline. The authors blamed a combination of agricultural water demand and drought, both of which, unfortunately, at that time had not been studied in any depth at all. 'The dramatic change' (Eimanifar & Mohebbi, 2007: 2) in Lake Urmia's decline was linked to '10 years of progressive dry climate in the area' (Eimanifar & Mohebbi, 2007: 2), and yet this statement was based solely on a master thesis by Azizi published in 2000, which may explain the oddity of a 2007 study referring to a study from the year 2000, after 'a decade-long drought' must have had occurred between 1990 and 2000, when the water level of the lake was at its peak. As such, the argument should be taken as a 'good guess' in retrospect rather than as a fact. Similarly, the human impact, which the authors extrapolated from other declining lakes in the USA and the growing population in the Lake Urmia basin, without any reliable data to support their argument—a recurring Neo-Malthusian theme, as we shall see—lacked scientific validity. The point here is not that the experts were ignorant or failed to warn of what was threatening the entire Lake Urmia basin but that the public or scientific debate took some time to register the abnormal variance in the water level; they most likely assumed that the lake's behaviour was within an acceptable norm, as Golabian (2011: 378) remarked:

> 'It was hoped that following the previous trends, the curve [(water level)] would turn upwards, but surprisingly it continued its downward trend and in the years 2009–2010 reached a historical bottom level'.

As the lake dropped, scientific inquiry on the origin of the crisis increased. The significant rise of published papers on Lake Urmia after 2012, however, was not merely based on scientific interest alone. The intersection between governmental politics, the academic community and the unveiling crisis of Lake Urmia were related in a specific interaction. In regard to the sciences, it would be a romanticised notion to believe that academic inquiry needs nothing but the immanent desire to conduct studies on a matter. Research also requires money and institutional support, and this reality is very much present in the academic world at Lake Urmia and Iran (Tamman, 2021). The increasing number of publications especially after 2012 is not some coincidence that could be explained by a strange, delayed sense of awareness on the part of the sciences but by a political move.

President Mohammed Khatami (1997–2005) and his successor Mahmoud Ahmadinejad (2005–2013) were driven by the paradigm of the hydraulic mission (Castelier, 2015), i.e., developing Iran primarily through large or mega-projects, such as the excessive construction of dams. Despite the lack of scientific work, the decline of Lake Urmia has been registered since the late 1990s and has since been on the political agenda (Hashemi, 2012: 124), albeit to varying degrees. According to Masoumeh Ebtekar, the former head of the Department of Environment (1997–2005), the first projects to save Lake Urmia were initiated in 2002. Three years later, though, with the inauguration of Ahmadinejad, these projects were suspended (Rajabova, 2013). In general, Ahmadinejad's presidency can be characterised by a mild disinterest in Lake Urmia—at least in the beginning. In 2011, when the decline was undeniable (water level down to around 1271 m and a surface area around 3400 km^2), Ahmadinejad normalised the phenomenon by insisting on the iterative nature of droughts that dominate Iran every 500 years, marking exogenous rather than local anthropogenic factors as the driving forces behind the lake's decline. Moreover, Ahmadinejad took his blame game even further, spatially speaking, by claiming that European countries had used technologies to modify the weather, i.e., Europeans deliberately took water from the clouds to deprive Middle Eastern countries of rainfall (Dalby & Moussavi, 2016: 15). All of Ahmadinejad's arguments implied that measures to save the lake would have little effect if the fault was not home-made (endogenous) but a deadly combination of a naturalised drought cycle and Europe's conspiracy against the Iranian nation (exogenous). This position called for inaction rather than initiatives to protect Lake Urmia. Nevertheless, during Ahmadinejad's late presidency, measures were taken to save the lake. In 2010, under the leadership of the Ministry of Environment, the Ministries of Agriculture, Energy and Interior and the governors of the three provinces of the Lake Urmia basin ratified a Memorandum of Understanding developed under the IWRM paradigm, which Iran had been trying to establish since the 1990s (MoU, 2010; Hashemi, 2012: 147). In addition, 20 MPs wrote a letter to Ahmadinejad urging him to take immediate action against the negative development of Lake Urmia;

however, measures to restore the water body could not be implemented due to a lack of budget (Salimi et al., 2019: 614).

In August 2011, protests erupted in the cities of Urmia and Tabriz, demonstrating dissatisfaction with the central government's lack of emphasis on the issue. Protesters chanted slogans such as 'Join us, Turkish brothers! Lake Urmia is in agony, Majles, [central government politicians] are signing the death warrant for the lake' (ADAP Azerbaijan, 2011). Although a local MP from Urmia, Javad Jahangirzadeh, called for a 'calm and logical atmosphere' to resolve the crisis, which should not be politicised, Lake Urmia did enter the centre stage of Iranian politics, so much so that in 2013, Ahmadinejad reportedly 'issued a special order to speed up efforts to restore [it]' (Azernews, 2013). The political awareness that Lake Urmia was a double-edged sword increased—since saving it could legitimise the central government, or, if it should vanish, portray it as a devastating failure and thus lead to increasing civil unrest in the north-western provinces.

Hassan Rouhani was conscious of the significant role the lake could play in either stabilising or destabilising the central government, and so as soon as he became the elected president in 2013, it became the centrepiece of the government's political agenda, whose objective was to tackle the country's looming environmental, and especially water, crisis. Rouhani's presidency marked a significant turn in environmental policies in Iran, whereby the *prima causa* of Lake Urmia's crisis shifted to endogenous or primarily local/anthropogenic factors. In 2013, Rouhani initiated restoration efforts and established the Urmia Lake Restoration National Committee (ULRNC), with the Urmia Lake Restoration Program (ULRP) as its executive unit (Salimi et al., 2019: 615). Subsequently, the number of peer-reviewed publications increased significantly compared to the Ahmadinejad period: of the 906 articles published up to 2022, 786 were published during the presidency of Rouhani and his successor Ebrahim Raisi (2021–2024). This suggests that the ULRP encouraged the scientific community to intensify its research efforts to study Lake Urmia (Parsinejad 2022: 11). Without political emphasis and funding, it is likely that less research would have been conducted in situ. The lack of financial support for academic research in Iran leaves scientists with two choices: to focus entirely on teaching or to find an interested foreign university that will commit to funding research projects on Lake Urmia as a study subject (Tamman, 2021). Under these circumstances, it is perhaps understandable that academic interest in

Lake Urmia has become a global phenomenon, since many papers are written by Iranians employed at foreign universities, as will be elaborated in the following chapter.

4.3 Scientific Landscape

The aforementioned MP Jahangirzadeh, in addition to calling for a calm, apolitical atmosphere—as if politics, either local and national, needed to be muzzled—added that the Lake Urmia crisis should be resolved endogenously, while 'foreigners have no right to air judgment on the issue' (Mehr News Agency, 2011). Yet, the opposite happened to be the case. Whereas in 2010 almost exclusively Iranian scientists engaged with Lake Urmia, researchers from 22 countries published five or more peer-reviewed articles by 2023, as visualised in Fig. 4.5. From 2013 on, various collaborations were formed between Iranian universities and foreign ones, international companies as well as government state agencies.

The United States, Germany, the Netherlands and Turkey are just a few to be mentioned. Moreover, in 2013, the United Nations Development Program (UNDP) and Japan reached out to the

Fig. 4.5 Network imagery of academia conducting research on Lake Urmia by 2023 (min. five peer-reviewed papers); own illustration with Gephi and data collected from dimensions.ai (accessed 20th June of 2023)

government to offer their assistance to increase restoration efforts (ULDP 2018: 143–144; Dalby & Moussavi, 2016). Four years later, the FAO (FAO & ULRP 2017) joined the list of fellow international organisations (Table 4.1).

Clearly, co-operation rather than isolation characterises the shared aspirations to save Lake Urmia, even at the cost of foreign condemnation and criticism of Iran's water governance. As Fig. 4.5 illustrates, a vast network of academic actors is involved in knowledge production relating to Lake Urmia.

Iranian universities continue to lead in terms of published papers, though. Of the 906 articles published up to 2022, 618 (68%) originate from Iran, 89 (9.8%) from the USA, 58 (6.4%) Germany and 46 (5%) Turkey. However, when looking at the specific universities conducting their research on Lake Urmia, the universities of Tehran, Tabriz and Urmia hold the majority of conducted studies, with 21.8% of documents and 18.3% of citations linked to these three universities alone. However, despite the seemingly equal distribution among them, most of the academic activity is located in Tehran, with 22% of published papers and 18% of all citations being traced back to the capital city alone. This underlines the central position Tehran takes in the debates surrounding Lake Urmia.

A second noteworthy observation is the number of technology-oriented universities writing and researching on the topic, which admittedly is not enough to be a valid insight, but if we add the research fields and journals to the preliminary observation, the initial assumption already expressed by Schmidt et al. (2021: 2149) can be confirmed. Data from the dimensions.ai search tool (accessed in June 2023) shows that the natural sciences dominate the scientific community concerning Lake Urmia, or less politically articulated, they contribute the great majority of studies. In fact, earth sciences take the lead with 476 publications, including hydrology (168 publications), atmospheric sciences (165) and physical geography and environmental sciences (147). Engineering (152) and environmental sciences (136) come in second and third place, respectively, while social sciences, with a total of 20 publications—8 of them by human geographers—are ranked in 8th place. The unbalanced distribution between social and natural sciences is therefore noticeable (data accessed via dimension.ai in July 2023).

However, the lopsided scientific engagement at Lake Urmia offers as many opportunities as it carries risks, one of which is the hegemonic

Table 4.1 ULRP Co-operations derived from ULRP (2018)

No	*Actor*	*Country*
Universities		
1	University of Twente	Netherlands
2	Center of Development Research Bonn	Germany
3	Utah State University	USA
4	Humboldt University	Germany
5	Aachen University	Germany
6	Wageningen University	Netherlands
7	Cornell University, New York	USA
8	University of California, Irvine	USA
9	The University of Melbourne	Australia
10	Tufts University; Boston	USA
11	University of Zurich	Switzerland
Regional and international organisations		
1	International Fund for saving the Aral Sea	International
2	International Center for Agricultural Research in the Dry Areas	International
3	International Institute for Applied Systems Analysis	International
4	Food and Agriculture Organisation	International
5	The Centre on Integrated Rural Development for Asia and the Pacific	International
6	Global Nature Fund	International
7	United Nations Development Programme	International
8	International Centre for Advanced Mediterranean Agronomic Studies	International
Companies		
1	German Water Partnership	Germany
2	Tayfur	Turkey
3	Koca Group	Turkey
4	ICT	South Korea
5	GeoFlow	Turkey
6	Fenno Water	Finland
7	SGI	Italy
8	Organiksa	Turkey
9	Richel Group	France
10	CREA	Italy
Foreign government agencies		
1	Ministry of Ecology, Energy, Sustainable Development & Spatial Planning	France
2	Wetskills	Netherlands
3	GTK	Finland
4	Embassies	International
5	Malaysian Agriculture Research and Development Institute	Malaysia
6	Japan International Co-operation Agency	Japan
7	Iran International Charity Organisation	International

ontological reductionism of water, and the other the—epistemologically speaking—objectification of human beings. In theory, the natural sciences—and I do not mean to disparage this point of view—believe in an objective reality, set in *one* world that can be discovered by positivist sciences (Venturini 2012: 267). This, 'mono-ontological' approach to water is by far the most dominant in the academic community—and thus in the water discourse on Lake Urmia.

Water is water, no more and no less. It is measured in m, m^2 and m^3, and it constitutes and is constituted by the hydrological cycle in which humans by default cannot do any good—they intervene, disturb, overuse, etc. Meanwhile, techno-managerial solutions are proposed as a simple calculation, a purely mathematical task: reduce demand, increase supply, problem solved! Nabavi (2018: 713) on this point notes:

> 'In Iran's academic and policymaking circles, engineering and mathematical knowledge prevails and humanities and indigenous knowledge are rarely taken into consideration. Non-engineering management options find less room to be fully considered. Instead, options stemming from command-and-control approaches and engineered solutions are viewed as reasonable policy actions'.

However, in recent years, socio-hydrologists have engaged with Lake Urmia in an effort to bridge the gap between natural and social sciences. In this context, the Iranian scientist Parsa Pouladi (from the Purdue University (USA)) is worth mentioning, as she elucidated the mostly neglected social processes in relation to the Lake Urmia hydrologic crisis, and thus the interrelation between the two (Pouladi et al., 2020, 2021). The socio-hydrologic movement has been received as a mixed bag by Madani and Shafiee-Jood (2020), as they criticise that the allegedly 'new science' labels old wine in new bottles. In general, Madani and Shafiee-Jood (2020: 17) warn that 'within the water resources profession', experts tend to stay in their self-designated comfort zone, 'read the same journals and go to the same conferences', which conserves and fosters their 'mental frames' but disregards other approaches that do not fit—through repetition—the established frame. The same goes for socio-hydrologists, as the community is still 'dominated by hydrologists' (Madani & Shafiee-Jood, 2020: 14). The lack of a social sciences perspective on the Lake Urmia crisis, as mentioned before, results in another problem, namely the objectification of humans, which is related to *modern water*. What people think of water might be interesting but nevertheless remains an 'imperfection of the human intellect' (Venturini 2012: 267). Worse still, people are reduced to actors who are incapable of learning, adapting and reflecting on their actions. Their actions remain a simple causality between Lake Urmia and themselves, and nothing else matters. People's thoughts, ideologies, values and positionalities are silenced by the thousands—with a few exceptions, as will be shown later.

The emphasis on engineering and technological sciences becomes even more self-evident by illustrating the most active journals and the bibliometric analysis publishing on Lake Urmia. It is undeniable that most of the journals, even if not explicitly mentioned, carry the odour of techno-managerial paradigms. The international journal *Water Resources Management*, which has published 31 papers on Lake Urmia, is a multidisciplinary forum supported by the European Water Resources Association, and its main objective is to support and promote collaboration between scientific knowledge and practical engineering activities. Accordingly, the journal's objective is to facilitate the management of water resources by addressing interrelated issues collectively (SpringerLink, 2024a). This gives the impression, to paraphrase Linton (2010: 215), that 'water is still a resource' and water management is still management fixated on controlling water. The second journal, *Theoretical and Applied Climatology*, essentially seeks to deepen understanding of the hydrologic cycle and the impact of climate change on this natural process (SpringerLink, 2024b). *The Arabian Journal of Geosciences* again focuses exclusively on non-human entities relevant to the natural sciences (SpringerLink, 2024c). The list of journals publishing on Lake Urmia continues almost exclusively in this way, which leads to the conclusion that most of the knowledge we gain from the lake revolves around—to use the old expression—'understanding objects and how to control them' or 'manage' them, to use the euphemism.

The dominance of the natural sciences and the water management paradigm is similarly observable in Fig. 4.6. The term co-occurrence map is based on bibliometric data from 906 peer-reviewed publications derived from Dimensions.ai's databank. The data was fed into the scientometric analysis tool VosViewer in order to extract and visualise the occurrence of and relation between the terms used in both titles and abstracts. Binary counting was applied, which indicates the number of documents in which a term occurs at least once. To avoid multiple

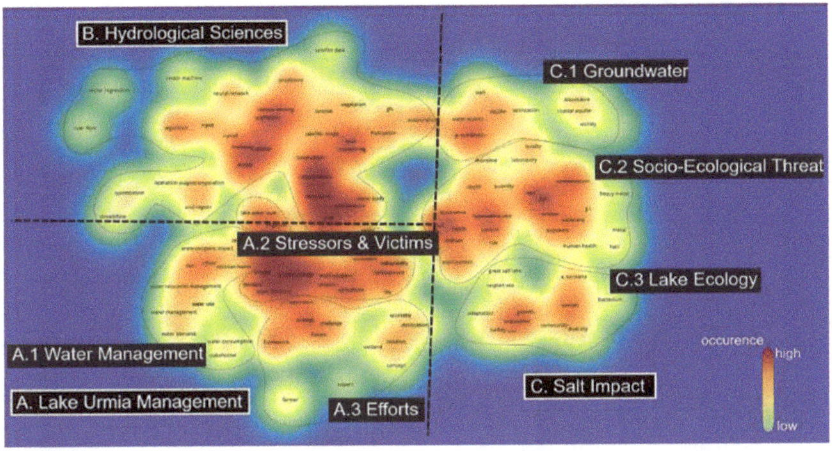

Fig. 4.6 Heat map of Lake Urmia sciences derived from Dimensions.ai and run through VOSViewer

occurrences of synonymous terms (e.g., manager and management) or grammatical variations (e.g., climate change and climatic change), a thesaurus file was created that merged terms based on similar topics to improve the legibility of Figs. 4.6, 4.7, 4.8. Of the 19,640 terms, only those appearing at least 10 times were used, and so 67 terms met the threshold.

The result of the following three co-occurrence maps helps gain a better picture of the sciences and topics surrounding Lake Urmia by focusing on the following characteristics: prominence, marginality, absence, relatedness and clustering. The purpose of the heat map in Fig. 4.6 is to show the 'topography' of the sciences, i.e., what aspects are covered by the research, and whether it is possible to identify coherent groups that fit into an overarching theme, hence a cluster. However, what gives each map its particular characteristics are 'structural holes, clusters, and sub-clusters' (Venturini & Munk, 2022: 194). Based on the interrelation of these holes and the cluster's density—partly disperse, partly concentrated—three main clusters can be outlined. The first, 'A. Lake Urmia Management', depicts topics that are both anthropogenic & managerial. Water here is a resource, managed, used, demanded and consumed (A.1 Water Management), whereas troublemakers (A.2 Stressors & Victims)

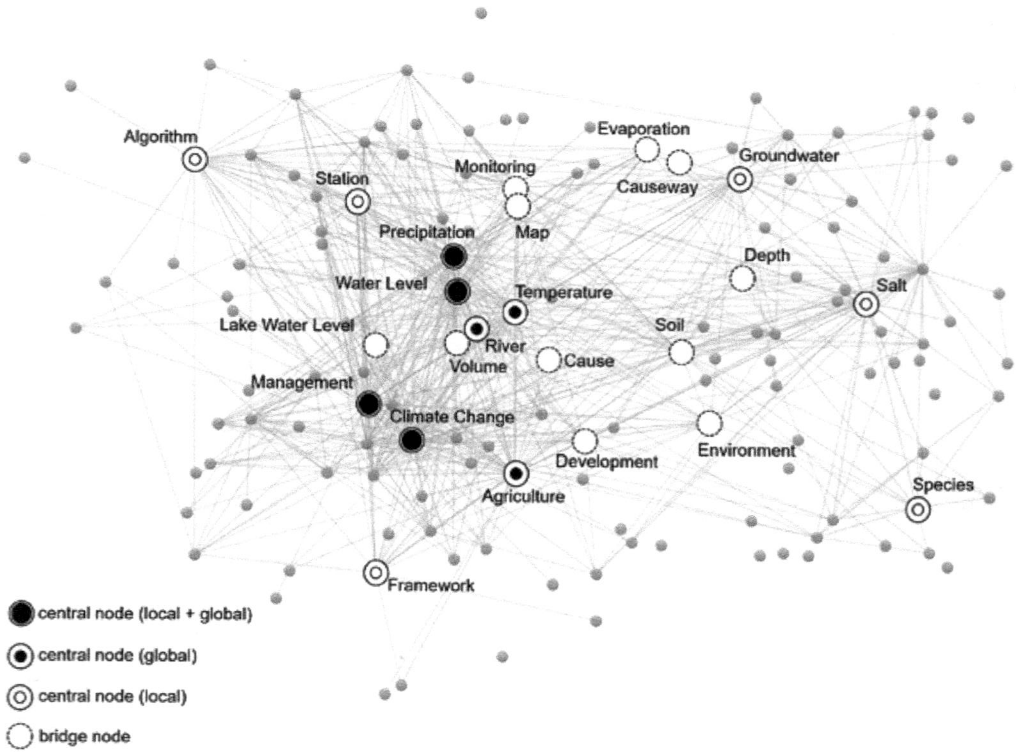

Fig. 4.7 Central and bridging nodes in the Lake Urmia sciences network (based on Dimensions.ai data and VOSViewer)

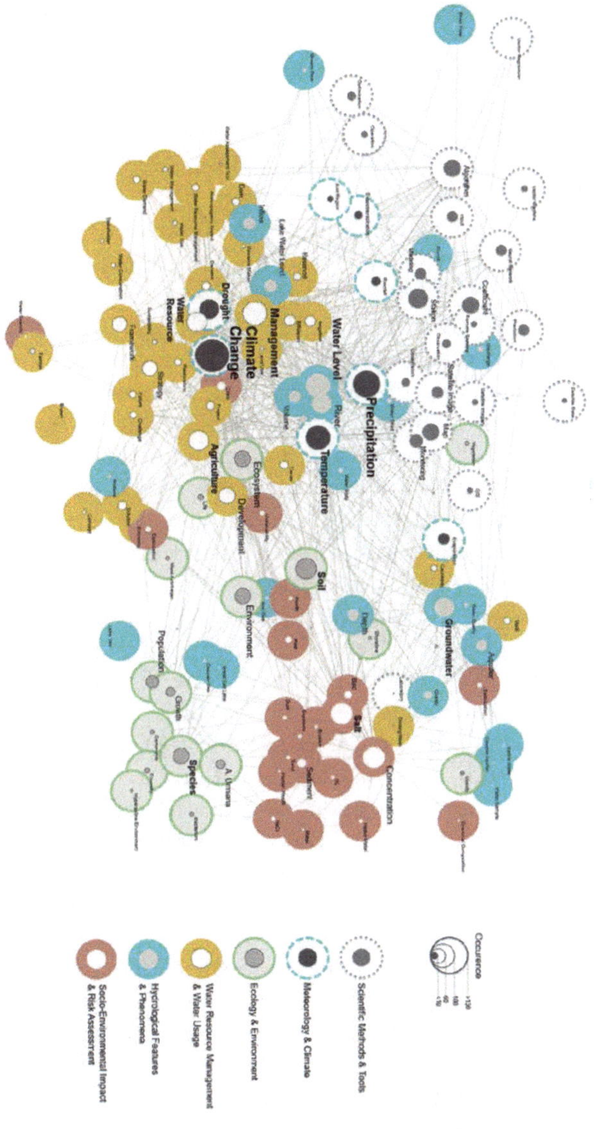

Fig. 4.8 Lake Urmia co-occurrence map of scientific papers

such as climate change, droughts, agriculture, bad water management, the anthropogenic impact and the crisis per se require (A.3 Efforts) strategies, restoration efforts, frameworks, development plants and solutions for the future.

Human actors, like farmers and experts, are also localised in this cluster. The second cluster, 'B. Hydrologic Sciences', describes the dominance of hydrological sciences in the matter of Lake Urmia. The water cycle, with a focus on precipitation, river flow and lake water levels, is assessed and predicted through models, algorithms, observations, satellite data and GIS. The third and last cluster emphasises the prominent role of salt as a field of interest in the sciences. Here, sub-clusters were applied due to the diversity of topics. The sub-cluster 'groundwater' could have been allocated to the hydrologic sciences, but due to its *closeness* to salt—and my knowledge from reading the literature—the relationship between these two is a more accurate representation of Lake Urmia. Scientists, as well as farmers, are concerned about salt intrusion into aquifers, which is particularly relevant in the vicinity of the lake. Overall, salt (dust) poses a threat both to the ecology, its diversity and survivability (e.g., *A. urmiana*) and to human society through its serious effects on physical health. This scenario is often associated with the aforementioned 'sister lakes', i.e., the Aral Sea (future dystopia) and the Great Salt Lake (fellow sufferer). The different 'heights' of the map, indicated by the number of occurrences of words, indicate the dominance of certain clusters. It is notable that cluster B. Hydrologic Sciences, and research on rivers, precipitation, water level and temperature, is dense next to the stressors and directly affected victims of the crisis in cluster A2. Cluster C is both less prominent and separated from the other clusters by a structural whole. Only sub-cluster C1, concerning groundwater, is loosely connected to the 'B, Hydrological-Sciences' cluster.

Figure 4.6 confirms the overall dominance of natural sciences and, more importantly, illustrates which terms are used most often. One is the analysis of stressors and 'victims'—alternatively, cause and effect—such as anthropogenic impacts, climate change, irrigation, agriculture and drought (stressors), and their negative effects on the ecosystem, agriculture, the lake water level and water resources (victims). The second focal point falls on the assessment of the hydrological conditions of Lake Urmia and the basin. In cluster B, the emphasis is put on measuring and quantifying the movement of water and the basin's—and thus Lake

Urmia's—water balance. These two aspects, if one imagines the heat map horizontally, are the largest mountains in this specific terrain, whereas the effect of salt, the lake's ecology and groundwater are comparatively hills.

The second map in Fig. 4.7 focuses on the positionality of specific actors or (sub) clusters, meaning which function they have in the network and how they relate to each other. The concept of a cartography of controversy draws upon the terminology of social network sciences, which identifies two key elements: centres, defined as local or global, and bridges (Venturini & Munk, 2022: 196). The challenge in this network is to determine the most central nodes. Those that have a central position in their cluster (local centrality) are Climate Change, Management and Framework (Cluster A); Algorithm, Station, Precipitation and Water Level (Cluster B) and the more distinct sub-cluster nodes Groundwater (sub-cluster C1.), Salt (sub-cluster C2.) and Species (sub-cluster C3.).

These nodes are locally central, meaning they play a pivotal role in their sub-cluster, whereas terms such as Temperature and Agriculture are globally central; they are embedded in a central position in the network, but they are not central in their respective sub-cluster. However, some of the aforementioned locally central nodes are also globally central, because their radiance or better, far-reaching connectivity to other sub-clusters in the network would completely distort the whole if they were left out. This implies that the left side of the network is the focal point of sciences at Lake Urmia. Management, Climate Change, Precipitation and Water Level are specifically central in the network as a whole. Bridge nodes, on the other hand, would similarly alter the structure, albeit not by changing the asymmetry of the network but by increasing the structural wholes. In other words, some nodes functioning to connect sub-clusters actually act as literal bridges, and without them, the network would crumble into smaller, unconnected parts. These nodes are Monitoring, Map, Evaporation, Causeway, Depth, Development, Environment and especially Soil, all of which connect Clusters A and B to C, which in turn implies that Clusters A and B are entangled to such a degree that the separation is somewhat arbitrary. Nevertheless, the nodes Volume and Lake Water Level do function as bridges between Clusters A and B, although their absence would only make a minor difference. Another way to present the network is to rank and categorise nodes by colouring and using their size to imply how often the term occurred in the scientific literature on Lake Urmia and which topics are dominant in the network (Venturini & Munk, 2022:

199). In contrast to the heat map, which focuses on the topology of the network and the function of the nodes in the map above, Fig. 4.8 emphasises the typology and prominence of nodes. This typology map helps to observe the dominance or marginality of certain nodes or categories. Finally, it can be compared with the clusters on the heat map, which allows for analysing whether they coincide territorially and in terms of content. The categories were chosen based on the thematic commonality of the terms. In addition, previous knowledge of the topic was used to find categories that are known to have effects on other actors (e.g., Salt). Six distinct categories were identified and coloured in this map: (i) 'Scientific Methods & Tools'; (ii) 'Meteorology & Climate', (iii) 'Environment & Geography', (iv) 'Water Resource Management & Water Use'; (v) 'Hydrologic Features & Phenomena' and (vi) 'Socio-Environmental Impact & Risk Assessments'. Furthermore, the nodes were ranked according to their frequency of occurrence, meaning that the more a term was used, the larger the node.

The typological map of the co-occurrence network helps to make several observations. First, certain nodes and clusters visibly dominate the network. Water management and water usage as a cluster are closely related to the most dominant nodes concerning meteorological and hydrological features such as 'Climate Change', 'Water Level', 'Precipitation' and 'Temperature'.

Second, the clusters applied in the heat map partially correspond to this map, but there are numerous nodes that spill over into other clusters. This is particularly the case for nodes from the category 'Hydrological Features & Phenomena'.

These nodes are scattered throughout the network and therefore cannot be reduced to a single location, although in terms of occurrence they are locally concentrated around certain nodes such as 'Water Level' and 'Groundwater'. Water in all its forms is spatially dispersed in the network, which marks it as the binding category, thus similarly diversifying the network as much as holding it together, in that water is in some way or another a bridge across every other cluster. Except for 'Scientific Methods & Tools' and 'Meteorological Features', other categories are intermingled with other clusters despite their local focus. This means that the clusters set in the heat map are not perfect but are still applicable enough to make a clear distinction; in other words, the 'visual clustering maintains a fuzziness that matches the ambiguity of collective phenomena' (Venturini & Munk, 2022: 205).

A third observation is—again—the clear focus in the sciences on hydrological, meteorological and water management issues, strongly supported by scientific tools and methods such as the use of Landsat satellite imagery/data or the application of algorithms and models to quantify hydrological and meteorological characteristics better. The proximity between 'Water Management' and the clusters above it implies, epistemologically, that the primary source of knowledge on Lake Urmia is derived from the natural sciences or quantitative data. Moreover, the water management category is the only one incorporating human actors—except for human health adjacent to 'Salt'. 'Farmers', 'Stakeholders', 'Decision-Makers' and 'Experts' are positioned next to solution-oriented nodes such as 'Framework', 'Sustainability' and 'Strategy', thereby implying that the scientific literature is also concerned with finding solutions or establishing pathways to escape the Lake Urmia 'Crisis'.

Fourth, the category 'Socio-Environmental Impacts & Risk Assessments' stands in contrast to other categories, as it is the only one based on a combination of interpretation and my knowledge of the literature. Evidently, other factors such as climate change or human activities also affect the environment. However, salt has a major impact on the lake's environment and ecology, as has already been demonstrated through the decline of A. urmiana—and more examples that will follow. Although other factors cause water scarcity, salt is one of the factors that has a serious and an immediate negative impact on both humans and non-humans, especially in relation to biodiversity, water quality, agriculture, groundwater, human and ecological health. Furthermore, Salt is a bridge node extending to scientific networks, so to speak, i.e., the Great Salt Lake, Aral Sea and Caspian Sea are only part of the scientific Lake Urmia network because their common denominator is salt—either because experts fear a dystopia comparable to the Aral Sea, try to derive solutions and similarities from the Great Salt Lake or propose to reduce Lake Urmia's salt concentration by using water from the Caspian Sea. This being said, it is not by coincidence that these 'lake' nodes are approximate to 'Salt' and its environmental impacts.

The other advantage this map offers is not only what can be seen, but also what cannot be seen. To grasp this notion better, it may be suitable to summarise what impression the scientific Lake Urmia network makes. Water is to be managed and made known by quantitative, natural sciences, and most of this epistemology is provided by a rather distant perspective. Satellites provide images and allow the construction of maps to monitor what is happening on the ground, while water is approached mainly through mathematical measures, which are fed by data from, among others, stations (e.g., hydrological stations). These entities, titled 'Scientific Tools and Methods' herein, enable researchers to provide references that ground their scientific endeavour and thus validate their work. Nevertheless, this grounding is primarily—or even exclusively—focused on non-humans: water, its movement and what drives or manipulates this movement, i.e., a world of objects colliding with each other causing all sorts of negative outcomes. Yet, what about the subjects, the human actors? Indeed, they are what is evidently missing in the network. Farmers, stakeholders and experts are visible, but there is little one can derive from these people living in the Lake Urmia basin, other than that they are part of a water management machinery. Humans are objectified in this case, stripped of their cultural or historical network—at least, that is what the map implies. To cite Latour (2005: 170):

> 'No one would dream of talking about [non-humans] without consulting the said [ones] by experimental procedures of unprecedented subtlety. With humans, though, we do not take so many precautions. On the pretext that humans are endowed with speech, politicians, like many survey specialists, sociologists, journalists, and statisticians, imagine that one can speak of them in their place and without ever truly consulting them — that is, without ever finding the risky experimental apparatus that would allow them to define their own problems themselves instead of simply answering the question asked. To parody Figaro: "In terms of the virtues that we require of objects, do you know many humans worthy of being non-humans?"'

As mentioned above, the majority of sciences conducting research on Lake Urmia stem from natural sciences, which are positivistic and based on a realist ontology, led by technical universities. Despite the presence of social sciences, and hence the existence of qualitative research methods, this research branch finds itself in a minority and is unable to traverse the hurdle to enter this map (the hurdle of having at least ten occurrences for a given term). This, again, implies that non-humans—from water, to meteorology and local species—is made known disproportionally more than humans. In addition, if humans are elucidated, it is mostly through statistical measures, namely quantitative methods again. To provide an anecdote from my field research, we were invited to join one of the ULRP's meetings in autumn 2018. When we asked the scientists whether they were including farmers' perspectives on the matter, they told us that they did so by uploading a 50-page-long questionnaire with closed questions. It may be a default position for natural scientists to be somewhat distant from their objects of interest; however, for them to maintain such a distance is by choice, since as humans we have the possibility to 'deep dive' (like ethnologists) into human worlds and their particularities.

This description of the Lake Urmia sciences network almost reads as a critique; nevertheless, it is not meant as such but to signal that there is a lack of perspectives on relevant human actors. Take, for example, the fact that politics, the ULRP, any of the Ministries, water governance, local protests, Lake Urmia's socio-ecological (you may say collective) history and many more entities are invisible in the network despite their prominent agency affecting the entire system. The human impact on Lake Urmia is veiled and generalised behind terms such as 'anthropogenic impact', 'water demand' or 'water use'. Nevertheless, the 'who', 'why' and 'how' exactly remain unknown. The scientific network around Lake Urmia lacks people, concrete political actors and a sense of history, and its social life—as it appears in the face of the networks above—is filled with lively objects and static, stubborn subjects: 'Human societies seem to be resigning themselves to playing the role of witless object, while it is nature that is unexpectedly taking on the role of active subject!' (Latour 2017: 74). As such, human actors should be granted more visibility in the network.

Despite the advantages of the co-occurrence map, it cannot show the trajectory or movement of the sciences. What can be observed, though, is simply which topics have been of primary interest so far—a superficial snapshot of the current scientific network. In order to support the argument that the majority of researchers at Lake Urmia focus on actors other than humans, Fig. 4.9 has been created. Its aim is to show which keywords have been used in the scientific papers—and thus which topics dominate the sciences over time and whether certain topics are gaining or losing momentum. With regard to Fig. 4.9, based on the 56 keywords collected during my analysis, five overarching themes were identified: hydrology, anthropogenic, climate, ecosphere and salt. Since 2014, the keywords #hydrologic (1st place) and #anthropogenic (2nd place) have led by some distance, while the keywords #climate and #ecosphere

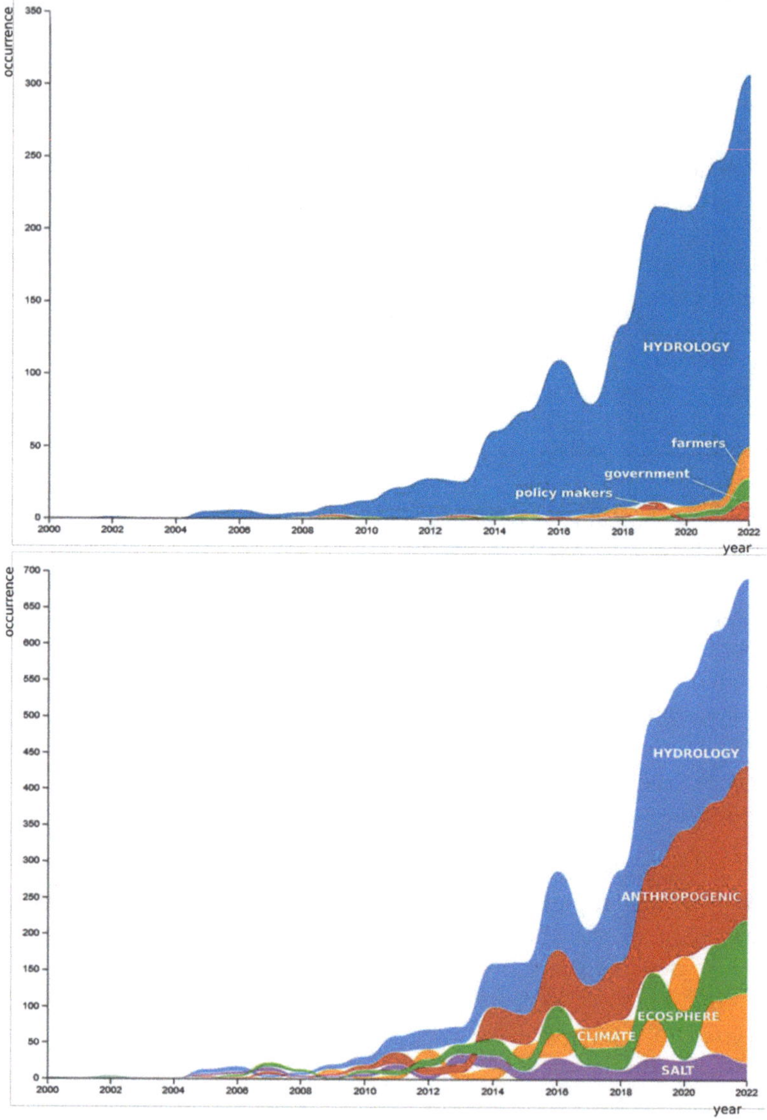

Fig. 4.9 Bumpchart (cumulative) of keywords (major categories, above; and "hydrology" keyword contrasted to human actors, below) used by scientific papers concerning Lake Urmia

are neck-and-neck. Salt never seemed to gain momentum, but at least it received steady attention. However, although the anthropogenic keywords are used frequently, they mostly refer to non-human actors being in an approximate relationship with humans, such as technologies that manipulate the flow of water (e.g., #infrastructure, #agriculture, #dam, #irrigation).

To illustrate the asymmetrical treatment of humans and non-humans in the sciences, Fig. 4.9 (second graphic) reveals how the keywords #farmers, #government and #policymakers are of marginal interest compared to, for example, hydrological keywords. However, since 2020, and especially since 2021, these 'human' keywords have gained momentum, receiving twice as much attention as in 2020. Nevertheless, they remain in a relatively understudied status quo.

As we show in detail in the following chapters, only a small number of scholars have recognised that the inclusion of people's thoughts, motives, concepts, traditions, stories, culture, habits and daily routines in the context of Lake Urmia has the potential to complement the findings and insights created by the natural sciences. Social sciences seek to reintegrate the people of Lake Urmia as relational, dialectical beings who reflect on their situation and decisions. In the scientific landscape of Lake Urmia, it is mainly non-humans who populate the hydro-social collective, while humans are—with some delay—only slowly introduced. As long as people are underrepresented in the sciences, the politics of Lake Urmia will miss any established target.

However, the first 'group' of actors that will receive the majority of attention in this study, as already mentioned, will be the sciences, or to be more specific, the scientific papers published on Lake Urmia. The reason for this decision is to highlight the fact that as long as Lake Urmia is a matter of concern rather than a matter of fact in every aspect, actors (as much as scientists) will discuss certain topics controversially.

Yes, they do agree upon certain things, and these shall be elucidated as well, but the goal here is to discover indifferences among actors. Controversies, one may suggest, are the driving forces behind Lake Urmia's hydro-social relations because the turmoil or 'uneasiness' they cause *makes* actors engage with topics in this matter of concern. Nevertheless, this does not imply that scientific facts or objectivity are just illusionary, social constructs but that matters of fact take time to evolve, and as we see in Fig. 4.4, scientific publications on Lake Urmia have only a short history, which has not allowed them to settle all differences. Sciences are a cornerstone of controversies, and thus they must be analysed to grasp the hydro-social metamorphosis, to understand what moves the actors other than physical objects.

4.4 Identifying Controversies and Relevant Scientific Actors

In the face of this task, a practical question arises: if I wish to analyse the sciences, how many papers do I have to read and integrate, and which ones are dispensable? In 2022, 906 papers were published with Lake Urmia as part of their abstract or title. Of these 906, roughly 660 could be considered a network in the sense that they relate to each other. In other words, papers that did not cite any of the relevant literature on Lake Urmia, and as such had no contact points to the network, were left out. Based on these 660 papers, Fig. 4.10 was created to serve a particular task: to identify those papers that must be analysed, which were optional and which one I could use to enter the analysis. In order to understand Fig. 4.10, the following points must be explained. First, the size of knots indicates link weight, i.e., how well embedded a certain article is within the network of literature. This shows either how often an article was cited or did cite articles from the community. This way, it is possible to literally see the authority of some articles. Accordingly, nodes that had a relatively high impact on the network and were substantial were chosen to be mandatory for the scientific discourse analysis (indicated by the blue nodes in Fig. 4.10). However, in this context the positionality of the dominant papers must be considered, because, again, this has important implications for the type of article. If an article is positioned within the hairball cluster, then it has been cited directly by others or cited by others frequently within the network. The more an article is located peripherally from the centre of the network, the less embedded it is. Take, for example, the following. The largest node, Hassanzadeh et al. (2012), is one of the most cited studies on Lake Urmia, even though it finds itself almost on the periphery of the hairball cluster. This is a logical consequence of the fact that this article was published in 2011, when the number of total published articles was significantly lower than after 2013. Hence, it could not have cited much of the literature that has become relevant over the years. Nonetheless, it plays a central role in the article network due to the numerous subsequent papers citing Hassanzadeh et al. (2012).

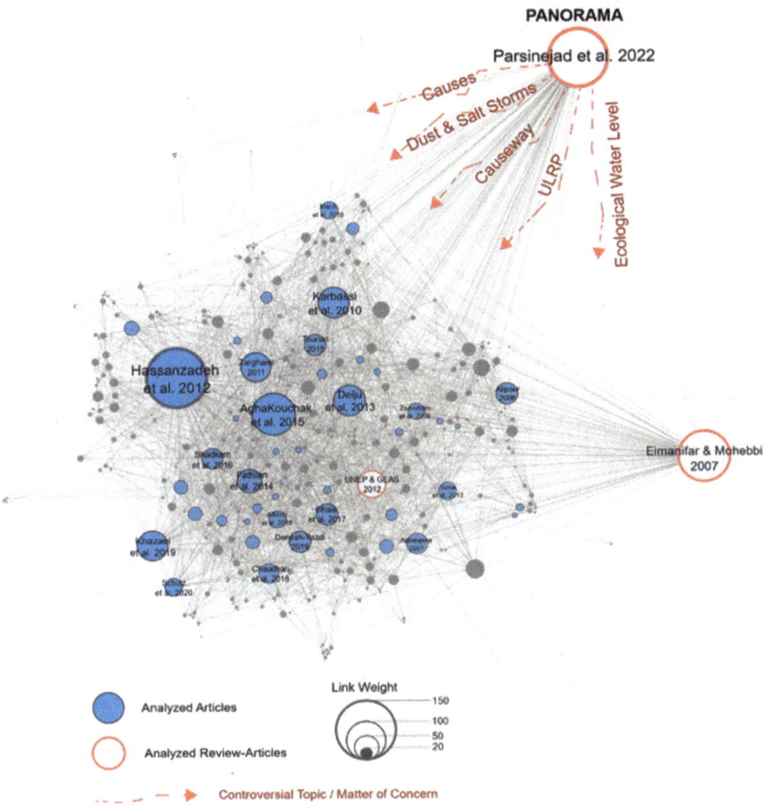

Fig. 4.10 Scientific article network on Lake Urmia

An even more distant, rather *exterior* position in relation to the hairball cluster is populated by the articles of Eimanifar and Mohebbi (2007) and Parsinejad et al., (2022). These articles, next to UNEP & GEAS 2012, function as review articles. For the purpose of this study, Parsinejad et al., (2022) was chosen as an entry point for the cartography of controversy, as it could be considered what Latour (2005: 187–190) designated a *panorama*.

Panoramas see everything, but in fact they do not see really anything at all, as they only 'show an image painted (or projected) on the tiny wall of a room fully *closed* to the outside' (Latour 2005: 187). Despite their limited nature, they offer the advantage of being able to preview the collective that follows—even though one should not confuse the collective assembled in a panorama with the real one in the common world. Panoramas 'design a picture which has no gap in it, giving the spectator the powerful impression of being fully immersed in the real world' (Latour, 2005: 188).

The synthesis report by Parsinejad et al. (2022) falls perfectly into one of the examples made by Latour (2005: 188), whereby a 'scientist summarizes for the benefit of the public "the present state of science"'. In a joint effort, researchers from the University of Tehran, Utah, Arizona and Istanbul co-authored and published in *Science of the Total Environment* their article titled '40 years of Lake Urmia restoration research: Review, synthesis and next steps'. Altogether, they reviewed 544 papers, 287of which made it to the final analysis—or synthesis. The number of reviewed papers explains why this document has the second-largest link strength in the academic literature on the lake. As Latour (2005) posited, panoramas are not *re*presentations of the scientific state of the art, in this case of Lake Urmia, but *pre*sentations. Politics and fiction do not halt before this panorama but traverse and alter it; against this backdrop, recipients should not mistake one scientific landscape drawn by some with the messy one that the cartography of controversy wishes to illustrate in this thesis, which is aware of its own politics of mapmaking (Venturini & Munk, 2022).

Thus, the study by Parsinejad et al. (2022) will help enter the controversies that lay ahead. Overall, five controversies are listed in a synthesis report that indicates the inconsistency of different scientific findings on certain topics. The first has probably gained some prominence in the sciences, as it reveals the heated debate on the main cause of Lake Urmia's desiccation, namely whether climate change or anthropogenic factors are more responsible for the crisis. This controversy has already been identified and shown how political the issue has become. The second incoherence Parsinejad et al. (2022) point out is the question of how dangerous Lake Urmia's aerosol emissions are, and if the lake is the sole source of dust storms. Moreover, and thirdly, even the effect of the causeway built across Lake Urmia is controversially discussed. Other controversies are the effectiveness of the ULRP and the ecological water target level set to restore Lake Urmia in sustainable manner. In other words, there are five 'gateways' through which one may enter the inquiry.

However, there is an important fallacy that must be avoided in the analysis and mapmaking of controversies. Although it has been possible to identify the most important documents representing the Lake Urmia discourse from the 'heart of the scientific network', the quality of the controversy maps will have suffered if only hegemonic perspectives were replicated—those most cited, renowned and, in the case of Iranian sciences perhaps, less politically critical. The aim of this thesis is to bring as many voices and perspectives onto the same playing field as possible, rather than reducing the complex common world to a single after-the-fact narrative. If sciences wish to contribute to a debate, and I side here with Venturini (2012), it is not complexity for which the researcher should strive but a map that displays a proportional tracing of public debate. Proportionality considers three different aspects in regard to viewpoints: (i) representativeness, i.e., the more actors share certain arguments or statements, the more visible they have to be made in the map; (ii) the more influence an actor can exercise, due to his position in a network, the more his viewpoints must be considered and (iii) granting space to dissenting minorities that might be otherwise overlooked despite their capacity to fuel a debate (Venturini, 2012: 3). So far, it is the latter point that has been missing, so it is important to include documents that enrich the number of perspectives on Lake Urmia from, for example, the social or political sciences—or at least studies that to some extent cover areas other than the natural sciences.

Yet, this pathway to assemble Lake Urmia's scientific associations does not suffice and runs the risk of still misrepresenting—in this case the scientific branch of—Lake Urmia's water parliament. Documents are important because their data, ideas, positionalities, values, solutions, dystopia and so on circulate and co-produce the perspectives of the sciences on Lake Urmia. Nevertheless, relying on them alone would distort the discourse into a highly academic treatise decoupled from socio-political processes. Less metaphorically speaking, it is necessary to know which documents have authority as much as which scientists have most of the authority. The difference between the two is as follows. A published study may have a certain prominence, but scientists in the flesh contribute equally—or perhaps even more—to the discourse around Lake Urmia. These scientists may have not published much on the subject, yet a co-authorship network elucidates their otherwise 'hidden' authority. Politicians, the public and the media are informed, on the one hand, by well-known studies, but above all by scientists *in persona*; the media, in particular, often turn to scientists to share their views on the lake. The great advantage of a co-author network analysis is the name-dropping of influential personalities from the academic community and beyond. In addition, the discourse receives a vitality that is sometimes missing in the quasi-objectively written academic papers, and it is way more difficult to depict in this more rigid and unemotional communication style. As soon as scientists turn to the public, the academic veil of a quasi-objective stance is.

often abruptly lost. In this case, these actors appear less as scientists and more as opinion-leaders of their cause, to which they are passionately committed. This in no way diminishes their academic achievements or the value or objectivity of their findings, but the articulation in the media becomes much more confrontational than is the case in more or less opinion-free studies. To identify the most prominent scientists that will be of relevance in this study, a co-authorship network was created as visible in Fig. 4.11. The size of the bubble indicates the number of total co-authorships, while the colour spectrum emphasises the average number of citations per author—from blue, to indicate a smaller number, to red, indicating a larger number of citations. The network helps to conclude some important characteristics, namely the authority of a few researchers and the centrality and impact of especially one network. First, the elephant in the room should be addressed. Madani Kaveh is perhaps the most prominent figure, and quite potentially the internationally most renowned expert, on the Iranian water crisis in general. Consequently, his dominant presence in this network may be less surprising; nevertheless, due to his influence, it will be necessary to give more insights on Madani's perspectives, values and solutions on Lake Urmia and Iran's water controversy in general. Other researchers who are additionally worth having a deeper look at are Tajrishy Massoud, Tourian Mohammad Javad and Fathian Farshad. These three researchers may not be as impactful in the scientific community as Madani, but they nevertheless occupy a central position in the network (e.g., Tajrishy is deputy head of the ULRP) or they act as bridges, which means that they are not so much embedded in a specific network but find themselves as central junction points between networks (e.g., Tourian & Madani).

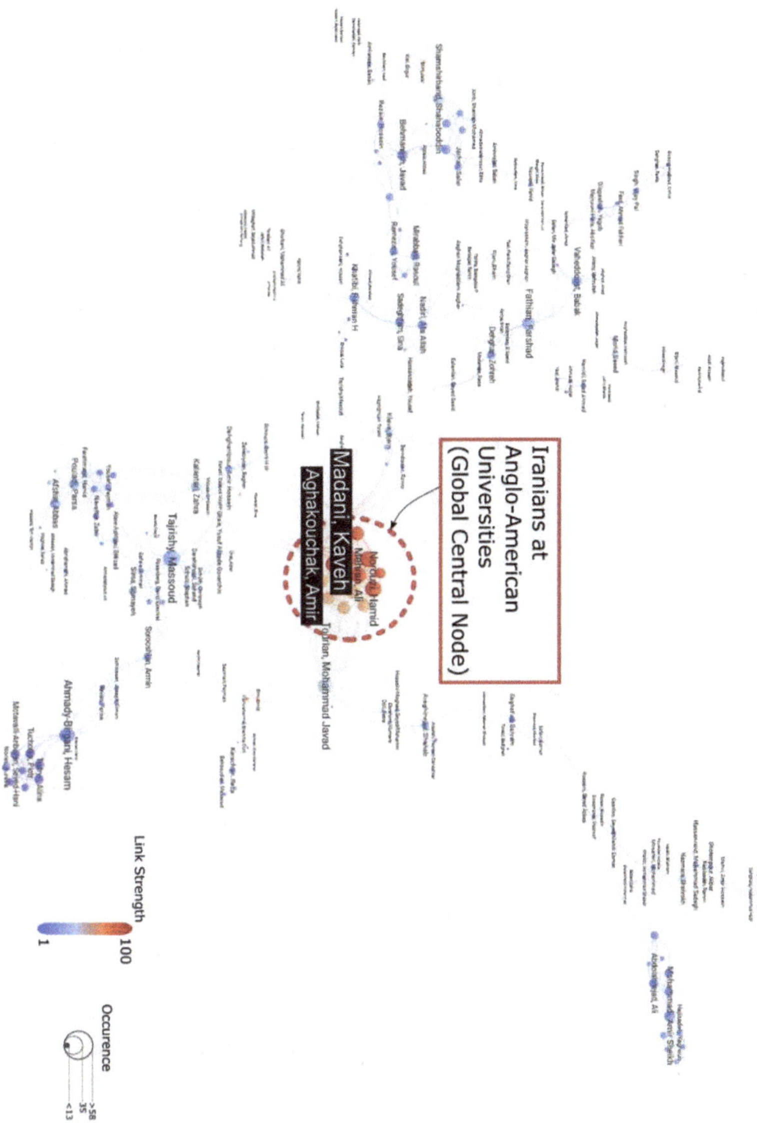

Fig. 4.11 Co-authorship network of Lake Urmia peer-reviewed publications

The second important observation of this network is the prominence or centrality of an entire cluster in the network. Specifically, Iranian researchers working at US-American colleges seem to have a sustainable influence on the scientific network, and they are thus by some distance the most cited authors in the entire network. This implies that it will be necessary to investigate how it transpires that this specific cluster—under the lead of Madani—has taken such a central position.

Conclusions: A Way Through the Invisible Nods The scientific community embraced Lake Urmia's diminishing fate especially after 2013, when President Hassan Rouhani established the ULRP, whereas beforehand, the lake received little attention from either central government or the sciences. Moreover, an increasing amount of international co-operation has developed ever since with numerous universities and organisations ranging from the UNDP, FAO, the Government of Japan and universities especially from North America and Western/Northern Europe. However, most of the scientific publications originate from Iranian scientists, who specialise in the techno-managerial and hydrological dimensions of the Lake Urmia crisis. As this chapter was able to show, narrowing down these factors has led to the neglect of social and political actors who are underrepresented in the scientific community. Although they do exist, they play but only a marginal role, at least in the scientific discourse. Other than the techno-managerial and hydrologic clusters that occupy most of the scientific attention, groundwater, salt and the lake's ecology have been revealed as important central actors as well. Keeping this in mind, the analysis of the scientific document network helped to establish a pathway that would allow one to follow specific controversies, i.e., topics on which scientists as well as other human actors do not seem to agree. According to the synthesis report provided by Parsinejad et al. (2022), which was identified as a panorama (Latour, 2005), five controversies offer themselves for further investigation in the following chapters: the causes of Lake Urmia's desiccation; the effect of the causeway built across Lake Urmia; the ULRP's effectiveness; the ecological water level and salt and dust storms.[1] Moreover, a scientific

[1] The "salt and dust storm" controversy is one I do not investigate further herein, since, and in contrast to the other four topics, it seemed to lack a sufficient body of work to elaborate on it sufficiently. Nevertheless, I would argue that is indeed a controversy worthy of study in the future.

cluster of Iranian scientists situated in the USA, and Kaveh Madani, as a prominent figure in the Iranian water crisis in general, can be highlighted as central actors in the scientific community, as they are highly active on the matter of concern, namely Lake Urmia's metamorphosis.

With this in mind, the following chapters will focus on the aforementioned controversies, paying attention to Madani and the ULRP as central actors in this matter.

References

Abbaspour, M., & Nazaridoust, A. (2007). Determination of environmental water requirements of Lake Urmia, Iran: An ecological approach. *International Journal of Environmental Studies, 64*(2), 161–169.

ADAP Azerbaijan. (2011). Tabriz protests for Save Lake Urmia (Sep. 03, 2011) تظاهرات تبریز. https://www.youtube.com/watch?v=GlVy3ktFzfQ&t=71s. Accessed 16 Augest, 2025.

Agh, N., van Stappen, G., Bossier, P., et al. (2008). Effects of salinity on survival, growth, reproductive and life span characteristics of Artemia populations from Urmia Lake and neighboring lagoons. *Pakistan Journal of Biological Sciences PJBS, 11*(2), 164–172.

Alesheikh, A. A. & Ghorbanali, A. (2004) Generation the coastline map for Urmia Lake by TM and ETM+ imagery. *Map Asia Conference.*

Azernews. (2013). *Iranian president orders speeding up efforts to restore Lake Urmia.* https://www.azernews.az/region/53500.html. Accessed 16 Augest, 2025.

Azernews. (2019). *Conditions created for cultivation of artemia in Iran's Lake Urmia.* https://www.azernews.az/region/150561.html. Accessed 16 Augest, 2025.

Bencherki, N. (2017). Actor-network theory. In C. R. Scott, J. R. Barker, T. Kuhn, J. Keyton, P. K. Turner, & L. K. Lewis (Eds.), *The international encyclopedia of organizational communication* (pp. 1–13). Wiley.

Castelier, S. (2015). *The demise of Lake Urmia sparks trouble in Iran.* https://www.middleeasteye.net/features/demise-lake-urmia-sparks-trouble-iran. Accessed 23 September, 2024.

Collins, H., & Pinch, T. (1993). *The Golem: What everyone should know about science.* Cambridge University Press.

Dalby, S., & Moussavi, Z. (2016). Environmental security, geopolitics and the case of Lake Urmia's disappearance. *Global Change, Peace & Security, 29*(1), 39–55.

Eimanifar, A., & Mohebbi, F. (2007). Urmia Lake (Northwest Iran): A brief review. *Saline Systems, 3*, 5.

Golabian, H. (2011). Urumia Lake: Hydro-ecological stabilization and permanence. In V. Badescu & R. B. Cathcart (Eds.), *Macro-engineering seawater in unique environments* (pp. 365–397). Springer.

Hashemi, M. (2012) *A social-technical assessment framework for integrated water resources management (IRWM) in Lake Urmia Basin, Iran.* Doctor of Philosophy Thesis. Newcastle University, Newcastle.

IRNA. (2022). Artemia a treasure deep in Urmia Lake. *IRNA English*, 2022. https://en.irna.ir/news/84614071/Artemia-a-treasure-deep-in-Urmia-Lake. Accessed 28 July, 2024.

Latour, B. (2005). *Reassembling the social: An introduction to actor-network-theory.* Oxford University Press.

Latour, B. (2013). Biography of an Inquiry: On a Book about Modes of Existence. *Social Studies of Science, 43*(2), 287–301.

Madani, K., & Shafiee-Jood, M. (2020). Socio-hydrology: A new understanding to unite or a new science to divide? *Water, 12*(7), 141.

Matthias, S. (2021). The vanishing of Iran's Lake Urmia. In: E. Ehlers, K. Amirpur (Eds.), *Middle East and North Africa: Climate, culture, and conflicts.*

Mehr News Agency. (2005). Economic news in brief (July 15). *Mehr News Agency*, 2005. https://en.mehrnews.com/news/12036/Economic-news-in-brief-July-15. Accessed Augest 3, 2024.

Mehr News Agency. (2011). 'Issue of Lake Urmia should not be politicized'. *Mehr News Agency*, 2011. https://en.mehrnews.com/news/47827/Issue-of-Lake-Urmia-should-not-be-politicized. Accessed Augest 3, 2024.

MoU. (2010). *Integrated Management Plan for Lake Urmia Basin.* https://www.ramsar.org/sites/default/files/documents/library/lakeurmiamanagementplan-i.r.iran2010.pdf. Accessed 16 Augest, 2025.

Nabavi, E. (2018). Failed policies, falling aquifers: Unpacking groundwater overabstraction in Iran. *Water Alternatives, 11*(3), 699–724.

Nimmo, R. (2011). Actor-network theory and methodology: Social research in a more-than-human world. *Methodological Innovations Online, 6*(3), 108–119.

Parsinejad, M., Rosenberg, D. E., Ghale, Y. A. G., et al. (2022). 40-years of Lake Urmia restoration research: Review, synthesis and next steps. *The Science of the Total Environment, 832*, 1.

Pouladi, P., Afshar, A., Molajou, A., & Afshar, M. H. (2020). Socio-hydrological framework for investigating farmers' activities affecting the shrinkage of Urmia Lake; hybrid data mining and agent-based modelling. *Hydrological Sciences Journal, 65*(8), 1249–1261.

Pouladi, P., Badiezadeh, S., Pouladi, M., et al. (2021). Interconnected governance and social barriers impeding the restoration process of Lake Urmia. *Journal of Hydrology, 598*, 1.

Rajabova, S. (2013). *Iran's official says water transfer to Lake Urmia unprofessional.* https://www.azernews.az/region/60438.html. Accessed Augest 11, 2024.

Salimi, J., Maknoon, R., & Meijerink, S. (2019). Designing institutions for watershed management: A case study of the Urmia Lake Restoration National Committee. *Water Alternatives, 12*(2), 609–635.

Sima, S., Rosenberg, D. E., Wurtsbaugh, W. A., Null, S. E., & Kettenring, K. M. (2021). Managing Lake Urmia, Iran for diverse restoration objectives: Moving beyond a uniform target lake level. *Journal of Hydrology: Regional Studies, 35*(9), Article 100812.

SpringerLink. (2024a). *Water Resources Management.* https://link.springer.com/journal/11269. Accessed Augest 29, 2024.

SpringerLink. (2024b). *Theoretical and Applied Climatology.* https://link.springer.com/journal/704. Accessed Augest 29, 2024.

SpringerLink. (2024c). *Arabian Journal of Geosciences.* https://link.springer.com/journal/12517. Accessed Augest 29, 2024.

Tamman. (2021). *The Hot List: Meet Kaveh Madani, the climate expert who fled Iran, April 27.* https://www.reuters.com/investigates/special-report/climate-change-scientists-madani/?utm_medium=Social&utm_source=twitter. Accessed December 5, 2022.

ULRP. (2018). *Urmia Lake: lessons and challenges.*

Venturini, T. (2012). Building on faults: How to represent controversies with digital methods. *Public Understanding of Science (Bristol, England), 21*(7), 796–812.

Venturini, T., & Munk, A. K. (2022). *Controversy mapping: A field guide.* Polity Press.

CHAPTER 5

Environmental Water Level and Causeway Controversy

The first controversy in this inquiry might not be the most prominent; however, it establishes an important baseline: What is the basic understanding of Lake Urmia? Why is it important? What is it made of, other than plain water? Which actors matter, and which less so? This chapter provides answers to these questions from the scientific perspective, given the fact that historicising and opposing different perspectives on the topic will deepen understanding on Lake Urmia. It will furthermore highlight how the lake is portrayed by different scientists. The first part of this chapter addresses a very specific topic: the minimal ecological water level of Lake Urmia, and in this regard, I demonstrate how the number has been calculated, on what basis and what it hopes to achieve. For this task, the tempo of this inquiry will be relatively slow, since it is vital to understand how the minimal ecological water level set at 1.274 m has come into being. Moreover, this calculation has not remained an isolated, scientific endeavour but has been enacted by law and turned into the single concrete target of the ULRP.

R. Gonda, *Lake Urmia's Hydro-Social Dilemma*,
https://doi.org/10.1007/978-3-032-08941-0_5

5.1 Lake Urmia's Health Level: The Shrimp-Salt Relation

Historically, Lake Urmia has received water from precipitation and river tributaries. However, the desiccation of the lake and growing concerns relating to an inevitable collective disaster prompted the Ministry of Energy and Water to deepen its understanding of the lake's *needs* (Hashemi, 2012: 180). Yekom Consulting Engineers is a private Iranian company established in Tehran in 1973 that has carried out numerous water-related projects, one of which was the development of an integrated water resources management plan for the Lake Urmia basin (Yekom n.d.). To assess Lake Urmia's water need, Yekom investigated the concept of ecological water based on Nazaridoust's doctoral thesis (Hashemi, 2012: 125–126). However, the scientific community came into contact with ecological water mainly through the co-authored paper by Abbaspour and Nazaridoust (2007) entitled 'Determination of environmental water requirements of Lake Urmia, Iran: an ecological approach'. The paper introduces, and as such frames, the subject in a Neo-Malthusian fashion: a growing population exploits water resources increasingly while availability is limited. This anthropogenic ignorance and lack of respect towards water bodies and their aquatic ecosystems culminates in an unbalanced ecology that threatens sustainable development. What is more, the authors even anthropomorphise water bodies as being capable of 'suffering' from mismanagement. Even if this expression should not be taken too literally, it still presents a particular state of mind, as we shall see. Abbaspour and Nazaridoust define the good water management of wetland ecosystems as including 'respect for all the activities within the wetland body as well as those throughout the associated catchment' (Abbaspour & Nazaridoust, 2007: 162). Respect for these water bodies—including river regimes—is expressed not only in the recognition of their value, but also in the definition of their environmental water requirements, which maintain their ecological function and thus ensure a sustainable environment. The 'experiment', as the authors called it, was to assess the amount of water necessary to provide a sustainable and ecologically healthy Lake Urmia. But even on this premise, certain ethical and ideological positions are visible.

The underlying assumptions of Abbaspour and Nazaridoust (2007) reveal three characteristics of their study: first, the ecocentric outline, second, the implicitness of nature as something vulnerable and third, nature as an external sphere contrasted with a social sphere. Ecocentrism expresses a purely environmentalist disposition. Contrary to ego- or homocentric positions, in which human interest and well-being, summarised as 'utilitarianism', are central to every decision-making process, non-humans placed under the umbrella term 'nature' are prioritised (Groenfeldt, 2019: 10). Water demand is assessed on the basis of the need for the ecological sustainability of water bodies and, as such, is assessed exclusively for non-humans. Moreover, the Neo-Malthusian framework imbues water bodies with the qualities of what Castree (2001: 6–7) describes as 'intrinsic nature', whereby external nature is fixed, pristine and predetermined by essential qualities. Water bodies and their aquatic ecosystems are intrinsically valuable to the environment but are vulnerable to exogenous, anthropogenic influences such as hydraulic infrastructure and increasing water demand (Abbaspour & Nazaridoust, 2007: 161). In this case, water is a contested but fragile object of interest between the environment and society. This vulnerability of nature has already been exposed as a dogmatic attitude that opposes nature as being able to heal itself quickly if guided by 'wise management policies' (Tvedt & Oestigaard, 2011: 4). *Nature ephemeral* expresses the vulnerability of nature, which diminishes its healing capacities or those of specific water bodies, e.g., river systems. On the other hand, ***nature benign*** emphasises the ability of water to heal itself. Both concepts, however, are dogmatic and represent nothing more than 'narrow conceptions of "nature"' (Tvedt & Oestigaard, 2011: 4). Nevertheless, Tvedt and Oestigaard (2011: 4) highlight the different aspects of water, on the one hand as something inherently capable of healing itself—falling from the sky as pure water, even without management strategies—and on the other hand as river systems in danger of being irreversibly degraded.

In essence, the introduction of Abbaspour and Nazaridoust's (2007) article attempts to explain, in a somewhat blunt manner, that water bodies, in this case wetlands, are ecologically important and must be acknowledged as such. Furthermore, a fix water requirement must be determined in order for the ecosystem to self-sustain. From an ethical perspective, such as that provided by Feitelson (2017), it is argued that ecological water needs are secondary to basic human water needs, but they are nevertheless fundamental to a functional and sustainable water hierarchy. Indeed, in a hydro-social collective, basic human needs must be met first; otherwise, the hydrologic regime is unsuitable for human beings to live in dignity. This is discussed, for example, by Feitelson (2012: 53–54) in 'What is water? A normative perspective'. Water for nature, which should also be normatively provided, is 'highly variable in space and time' (Feitelson 2012: 55), and a distinction must be made between minimum needs and the 'full array of environmental services' (Feitelson 2012: 56), as the former must be balanced, or rather negotiated, with other water demands.

Abbaspour and Nazaridoust (2007) essentially try to calculate a specific amount of water that allows for a certain type of aquatic system to persist and continue its habits, i.e., its usual functions. The premise of this paper is not necessarily based on 'hard facts' but primarily on an ideological stance, which promotes (to some degree) modern water. In this sense, water and its ecosystem are portrayed as something precious but limited while simultaneously being threatened by human intervention, which marks it as fragile. This move is not unscientific per se, but to a certain extent, it makes assumptions that are still contested in the sciences, such as the Neo-Malthusian perspective or the universal vulnerability of aquatic systems. The positionality is, so to speak, a choice, but it is necessary to legitimise the necessity to define an existential minimum for the lake. Lake Urmia, its wetlands and its catchment area are treated as something to be politically included, which means that their value to human society should be recognised and included but nevertheless still strictly differentiated. The calculated water requirement is exclusive to Lake Urmia and its ecosystem, which, however, drives the ontological setting into two distinct spheres: a quantity of water that is assessed without taking into account the human need for water, and portraying Lake Urmia and its waters as a protected area that should be left untouched by anthropogenic factors as much as possible. In doing so, the authors draw two circles—one for nature and one for humankind—which renders them antithetical. The continuity of ecological functions and the persistence of a specific type of Lake Urmia is anchored by a specific number, which, in other words, assigns a quantity of water to guarantee an existential aquatic ecosystem minimum.

Against this backdrop, Abbaspour and Nazaridoust (2007: 163) define their main objective as '[finding] the most critical ecological element(s) of the ecosystem and then [regulating] water demand so that the element(s) may have a sustainable status'. In order to do so, they base their calculation on a step by step process, which includes the following: (i) describe the essential environmental setting relevant for this experiment; (ii) determine ecological objective(s) as well as related indicator(s) based on Lake Urmia's environmental inventory; (iii) identify and aggregate water quantity and quality needs of the indicator(s) and (iv) calculate water requirement that meets hydrological, ecological and biological needs, in order to sustain the ecological characteristics of Lake Urmia.

As a first step, the authors decided to describe the Lake Urmia ecosystem as a combination of its hydrologic conditions Keltwater quality and water quantity) and its ecology (flora and fauna). The reference chosen for water quantity was the water level in the Persian Gulf, measured at CD (chart date). The indicator for water quality was set at the proportionally most dominant chemical component, NaCl, which accounts for 90% of total dissolved elements in the lake. However, although other studies (e.g., Kelts & Sharabi, 1986: 112) confirm the dominance of NaCl, Abbaspour and Nazaridoust (2007: 164) failed to provide a legible pie chart for the lake. Coupled with the lack of references for the data—provided by the West Azerbaijan Regional Water Expert Committee—readers of their paper will either trust or distrust the figures presented therein. Furthermore, the authors violated or ignored their self-omitted process because they determined the indicators before they described the Lake Urmia environment, which is problematic, as the relationship between the components is at first not evaluated properly; it is unknown why water levels were chosen as a quantitative indicator, or why NaCl is the sole determinant of water quality in Lake Urmia, because no correlation has been established other than a habitual assessment of water quantity and NaCl as the dominant factor. It is only in the next step, which should have been addressed beforehand, that the ecosystem is described. The authors summarise Lake Urmia's ecology as 'very sensitive and ecologically vulnerable' due to its 'extremely simple ecological pyramid', for which the brine shrimp *A. urmiana*—one of 12 species in the lake—lays the ecological foundation (Abbaspour & Nazaridoust, 2007: 164). As *Artemia* is the only food source for migratory birds such as flamingos, any change in its population harms the ecosystem. Other species, such as algae or bacteria, were not considered but only indicated via a slightly cryptic explanation:

> 'Thus, the fauna in the aquatic section of Lake Urmia is limited to A. *urmiana*, the status of which at the top of the simple ecological pyramid of the lake could show the lower levels of the aquatic ecosystem (i.e.,

producers). Hence, *A. urmiana* is selected as an ecological indicator of the ecosystem'. (Abbaspour & Nazaridoust, 2007: 165)

In other words, what the authors may have intended to say was that a healthy population of *A. urmiana* as an ecological indicator equates to an overall healthy aquatic ecosystem. *A. urmiana* thus takes on the role of representing the health level of any other flora and fauna in Lake Urmia, without the authors feeling the need to describe or establish any relationship with these other actors. Terrestrial fauna is reduced to an indirect relationship to the lake, whereby the fauna—especially migratory birds such as the flamingos—only feed off the shrimp or use the lake as a place for breeding or shelter. *Artemia* thus sits at the top—or as the basis, depending on the perspective—of Lake Urmia's shallow eco-pyramid. It is described as both fundamental and the most important indicator and thus gains a prominent status. The lake's vulnerability thus stems from *A. urmiana*'s sensitivity. The more the shrimp experiences degradation, so does the lake—especially as an ecosystem and perhaps less as a hydrological body.

The water quality and quantity indices were briefly addressed by Abbaspour and Nazaridoust (2007: 165). The two essential insights were that the mean Lake Urmia water level from 1965–2005 was at 1275.86 m, while NaCl concentration (in parts per thousand; ppt)—in the words of the authors—'has not been steady', or more specifically, inversely related to the water level, which means the higher the water level, the lower the NaCl concentration. This relationship between the water level and NaCl concentration sets the basis for the third component in this calculation. *A. urmiana*, the centre piece in terms of ecological health, is directly correlated to NaCl concentration. However, this relationship is more nuanced than that of the water quantity and water quality indicators. Here, the authors attempt to identify a particular NaCl concentration range that allows *A. urmiana* to survive. Therefore, they declare a minimum NaCl concentration, as well as an optimum and a maximum. The extremes are set right at the beginning, whereby *Artemia* needs salinity higher than 70 ppt and cannot survive salinity higher than 320 ppt. Within these two set parameters, the authors discuss the optimum conditions for *Artemia*, assessing that laboratory conditions may prove that its lifespan decreases as salinity increases, albeit the shrimp seemed to be more vulnerable in laboratory conditions where salinity concentrations of 75, 125 and 175 ppt were applied. In their natural habitat, *Artemia* populations appeared to be stable even with salinity higher than 200 ppt. However, even though *Artemia* could survive a NaCl concentration at 300 ppt, population growth would 'be very low', while 'most of the physiological activities [would] cease' (Abbaspour & Nazaridoust, 2007: 166). Moreover, at 250 ppt, *Artemia* cysts seem unable to hatch, which marks this concentration level as the absolute threshold for its sustainability. The authors, based on these assessments, decided that salinity should not exceed 240 ppt; and I mean intentionally 'decided' because none of this was calculated but founded on other references. Recipients of the study therefore cannot recreate the reason why the maximum threshold level was set at 240 ppt other than to trust the researchers' own conviction. Given this ill-founded approach, Parsinejad et al. (2022: 13) commented that 'the "Ecological Level" is based only on anecdotal information'. Based on the 240ppt NaCl concentration, a water requirement was calculated, with results showing that the minimum ecological water target level is 1,274.1 m, which equates to a surface area of 4,652.2 km^2 and a required water input of 3,086 Mm^3. Moreover, the researchers provided a satellite image of Lake Urmia's extent, if this target level is reached. In this scenario, the lake would be in hydrological and ecological equilibrium, while the water would even cover 'the crystallized salt zone' (Abbaspour & Nazaridoust, 2007: 167). The figure provided by the researchers, however, shows a calculated elevation of Lake Urmia that has been reached—in comparison to other satellite images provided, for example by the ULRP (2015: 2) during the late 1990s and 1980s, when water levels were at their peak. This is probably best observable with the largest in-lake island, Islami Island. In the figure provided by Abbaspour and Nazaridoust (2007), the island is visibly cut off from the mainland and has a surface area of 4,652.2 km^2. However, in satellite images used by the ULRP, Islami Island was connected to the mainland in 1991 and spread across an area of 4,858 km^2, therefore making it a peninsula. Only at an area of 5,249 km^2 could the island be distinguished from the mainland with the naked eye.

Altogether, the study by Abbaspour and Nazaridoust (2007) partly calculated and estimated the minimal ecological water level. The outline presented a vulnerable aquatic water system that is facing an inevitable scarcity of water due to the growing demand for water by human actors. The sustainability of Lake Urmia's aquatic system, and as such of many non-human actors directly or indirectly correlated to the lake, was prioritised and therefore opposed and separated into human water needs—certainly not intentionally but simply by choosing an ecocentric position. In other words, the main objective of the researchers was to define a water level that would restore and maintain Lake Urmia for the sake of the lake itself. The assessment of an ecological water level is based on the relationship between the water level of the lake, its salinity (expressed in NaCl concentration) and the local brine shrimp *A. urmiana*. Based on this assessment, a minimum target water level was posited to achieve ecological equilibrium. From a hydrological point of view, Abbaspour and Nazaridoust (2007) were particularly interested in defining a specific chemical constellation of Lake Urmia water that gives *A. urmiana* the potency to sustain its population. Put more simply, water's main quality was its function as a habitat for biotic non-humans. The quality of Lake Urmia would be adversely affected by water with a concentration below 70 ppt and above 240 ppt. The water's capacity to nourish the surrounding environment would be diminished, potentially leading to the deterioration of the lake's ecological balance. As a result, two types of water can be defined: one that falls within a certain range (70 ppt–240 ppt) and serves to support life, and the other, relevant in this case, that is incapable of sustaining the Lake Urmia environment (salinity above 240 ppt) and may even be destructive (above 320 ppt). This helps define, roughly speaking, three different types of water. The first is too low in saline concentration and thus does not sustain *A. urmiana* or, supposedly, the entire lake's ecology. The second is too high in saline concentration. And lastly, the third is a spectrum in between these two hostile extremes where *A. urmiana* can sustainably exist and where within these borders lies an incalculable optimum for its greatest reproduction rate.

The reason why I have paid so much attention to this study is that the ecological water level of 1,274.1 m was chosen by the ULRP (2015; 2018) as the main target level, the goal around which the whole restoration effort revolves. However, the ULRP did not choose this figure by chance but was de facto forced to do so. In 2008, the concept of an ecological water minimum for Lake Urmia was incorporated, signed and endorsed by stakeholders in a Memorandum of Understanding (2010: 24), approved by the High-Water Council's technical committee in 2009 and, finally, one year later, approved as a by-law by the Council of Ministers (Hashemi, 2012: 184). The once scientific endeavour to define a minimum ecological water requirement for Lake Urmia, now enacted by law, means that managers, companies, ministries and institutes align every action to save Lake Urmia in accordance with the water level of 1,274.1 m.

5.2 The Ecological Water Level is 'No Magic Number'

The ecological water level, however, has not remained uncontested. In recent years, multiple studies have looked for alternative indicators to assess the success of restoration efforts. In considering the panorama once more, it becomes evident that two distinct schools of thought have emerged within the scientific community with regard to the evaluation of Lake Urmia's ecosystem services. This is evidenced by the meta-study conducted by Parsinejad et al. (2022). The one discussed above has been labelled as a 'single-level' target, while its critics wish to implement 'multi-level' lake assessments (Parsinejad et al., 2022: 9). Apart from this binary opposition, most scientists support the former, namely the ecological water level of 1,274.1 m. Parsinejad et al. (2022: 11) complain about the fact that '[n]early all studies on lake hydrology and water management uncritically treated the "ecological level"', while '[t]he target lake level may not maintain important lake functions'. Interestingly, when I asked Parsinejad et al. (2022) to send me their paper, they also attached Sima et al. (2021) and their study on 'Managing Lake Urmia, Iran for diverse restoration objectives: Moving beyond a uniform target lake level'. The two driving forces behind the push towards level variability and multiple indicators for Lake Urmia's ecological status are Somayeh Sima and David E. Rosenberg.

Sima, nowadays working at the Faculty of Civil & Environmental Engineering at Tarbiat Modares University in Tehran, travelled to Utah in 2018 on a visiting fellowship, where she met Rosenberg, a professor at the Department of Civil and Environmental Engineering and the Utah Water Research Laboratory. Both researchers addressed single water level targets critically, as they stated in an interview. Rosenberg, for example, expressed, '[w]e can't say that restoring the lake to *some magic number* will improve the overall situation. Instead, we need to consider how [its] ecosystem services are interconnected and how a varying lake level will impact those systems over time' (Holman, 2021; emphasis added). Sima therefore concluded '[w]e have to embrace lake level variability and focus our restoration efforts where it makes sense' (Holman, 2021). To relate to an ecological water balance, which—as has been shown—was constructed by another group of scientists as 'some magic number' raises the question about how well constructed and 'factual' the proposed number of 1,274.1 m really is. A priori, that is without having looked at their study (Sima et al., 2021), we can identify two distinct groups (single-level and multi-level) and their main differentiator. As Rosenberg contested, the ecological water balance wishes to normalise a fixed water level while, as Sima added, the second group wishes to normalise the *variability* of Lake Urmia's water level. Thus, the former wishes to increase the lake's water level at all costs, as it is the *only* indicator, while the latter diversifies the 'simple ecological pyramid' and in a political move stretches the radius of their circle, i.e., the one that will now include even more factors other than water levels, salinity and *A. urmiana*. In their co-published paper, Sima et al. (2021) advocated multi-level targets as opposed to a single-level target.

The study differs significantly from the outset in terms of ethical positionality. In contrast to the ecocentric evaluation of Lake Urmia by the proponents of a single-target water level, Sima et al. (2021) frame their work in a hybrid mix of anthropocentric and ecocentric ethics—a position Merchant (2017) coined with the term 'partnership ethics'. This means that the value of Lake Urmia goes beyond its ecological self-worth. Moreover, the lake has social and economic value through its ecosystem services, and these services need to be identified, measured and their 'well-being' optimised through the evaluation of specific indicators. Thus, Sima et al. (2021: 3; emphasis added) define the components of a successful lake restoration as follows: '[It] must define objects, quantify metrics to indicate when objectives are reached, and integrate across connected *natural and human* components from the contributing basin'. Moreover, the authors warn that a single lake level cannot guarantee that biotic and abiotic actors will benefit equally as a result. They therefore call for the implementation of indicators that take into account the different ecosystem services. These indicators consist of eight objectives under the categories of water quality, ecology, human health and recreation.

The objectives defined by Sima et al. (2021: 4) are to maintain salinity below saturation (water quality), to conserve *A. urmiana*, bird species and endemic mammals (ecology), each with its own indicator, to reduce dust from the dried lake bed, to maintain the economic value of $Mg2 +$ and to restore ecotourism. Water quality was assessed differently by Sima et al. (2021) than by Abbaspour and Nazaridoust (2007). First, salinity is not only indicated by NaCl concentrations, i.e., salinity consists of more chemical components than just NaCl. Therefore, Sima et al. (2021: 3) clarify that an NaCl concentration of 240 ppt corresponds to a salinity value of 263 ppt.

This new number implies how distorted the focus on NaCl may have been and how the incorporation of *total dissolved solids* (TDSs) changes the perspective on the issue. The complication for a reader that results from this different understanding of salinity causes confusion in terms of whether both research teams are talking about the same 'salinity'. For example, in Abbaspour and Nazaridoust (2007: 165–166), NaCl saturation at 320 ppt is mentioned once as the absolute threshold for *A. urmiana*. Thereafter, *salinity* is never addressed again, either as NaCl concentration or in any other unit, and it is never expressed in TDS. Therefore, every time the authors mention salinity, recipients automatically think of NaCl concentrations. The cognitive correlation becomes even more viable as the authors conclude that salinity higher than 250 ppt would drastically prohibit *A. urmiana* from flourishing. This was the reason why recipients might have silently nodded at the conclusion that 'the salinity concentration of 240 ppt of NaCl is considered as the water quality requirement for the survival of *A. urmiana*' (Abbaspour & Nazaridoust, 2007: 166). To add to this confusion, Abbaspour and Nazaridoust (2007: 166) set the NaCl threshold for *Artemia* at 240 ppt because they referred to a study that suggested that salinity higher than 250 ppt would stop *Artemia* cysts from hatching. But what concentration does the 250 ppt replicate? NaCl salinity or TDS? Obviously NaCl, because if you set the salinity of NaCl at 240, that would be 263 ppt TDS, but 263 ppt would be 13 ppt above the corresponding threshold of 250 ppt. To give an example of how confusing the whole matter is, one can look at Hashemi's PhD thesis, in which he mentions that Abbaspour and Nazaridoust (2007) considered salinity as the main indicator of water quality because *A. urmiana* is particularly sensitive to sodium chloride (NaCl). Moreover, Hashemi (2012: 181) writes that 'the optimal salinity for normal reproduction of *Artemia* is 240 [ppt]', although NaCl was considered as the water quality indicator as a proxy for salinity and the 240 ppt corresponds to NaCl concentration, *not* salinity concentration. This might explain why Sima et al. (2021: 3) felt the urge to correct the NaCl concentration at 240 ppt as equivalent to a salinity of 263 ppt, expressed as TDS. This latter number was to be found nowhere else—in neither other studies nor in official reports.

Abbaspour and Nazaridoust (2007) and Sima et al. (2021), however, agree on the fact that bioassay experiments in the laboratory with *A. urmiana* do not simulate the brine shrimp's natural habitat, and therefore the observed optimum of salinity at 75–100 ppt (Agh et al., 2008: 169) cannot be used when assessing ecological water balance. However, the point at which the optimum level—or at least the threshold—for *Artemia*'s sustainable growth lies is a highly contentious issue, as there is a lack of data on *Artemia* populations at salinities between 170 and 235 ppt (Sima et al., 2021: 6); concentration levels for pre-crisis Lake Urmia fluctuated between 140 and 220 ppt prior to 1999 (Agh et al., 2008: 164).

Even comparisons with other *Artemia* species such as *Artemia franciscana* cannot be used to calculate a threshold for salinity concentration, as *A. urmiana* is more resistant to high salinities (Sima et al., 2021: 6). In addition, because data is limited, comparisons to other lakes and *Artemia* not being transferable and 'rapid evolutionary responses' (Sima et al. 2021: 8) not being taken into account, an exact salinity concentration cannot be established. What is certain, though, is that the higher the salinity, the lower the *A. urmiana* population density. However, exactly where the threshold lies remains unknown, and as explained above, it remains a controversial figure. Therefore, Sima et al. (2021: 8) concluded that the target level of 1,274.1 m (263 ppt salinity) would not be sufficient to produce an *Artemia* population capable of feeding larger numbers of flamingos and other birds. On the contrary, the ecological water balance could be detrimental to *A. urmiana*.

Moreover, another factor that has been neglected in the ecological water balance is the lake's changing bathymetry due to salt layers stacking up on the lake's bottom, which leave a static target water level obsolete (Sima et al., 2021: 12). In summary, Sima et al. (2021) are convinced that the abiotic-biotic relationship between the lake as a hydrologic object, migratory birds such as flamingos and brine shrimp *A. urmiana*, as well as other invertebrates, cannot be subsumed to a single-target level of 1,274.1 m. In their own words (Sima et al., 2021: 12):

> 'In the face of these variations and uncertainties, a uniform lake target such as 1274 m cannot also simultaneously achieve multiple desired ecosystem services such as reducing the salinity to 263 g L^{-1}, increasing *Artemia* densities, and increasing flamingo abundance. In fact, no single lake level can simultaneously satisfy all management objectives'.

Other studies that have touched on the ecological water balance at 1,274.1 m may not have been as overtly critical, but their scientific findings still reveal weaknesses in this target level. For example, according to Alborzi et al. (2018: 7), the ecological water level fails to include dynamic climate changes, and the static ecological water demand of 3,100 Mm^3 for Lake Urmia may be incapable of coping with water stress. The drought period from 1998 to 2002 stopped the lake from rebounding to an ecological threshold, and so in the aftermath, the lake level oscillated at an average of 1,273.5 m. The additional anthropogenic withdrawals that followed under near-normal climatic conditions put more pressure on the lake, further reducing its resilience. In conclusion, Alborzi et al. (2018: 1) propose a dynamic assessment of the environmental flow requirements of Lake Urmia. Accordingly, the lake requires 2,900 Mm^3 of water per year in dry conditions and 5,200 Mm^3 in wet conditions. This would correspond to an average environmental inflow of 4,100 Mm^3, which is 1,000 Mm^3 above the fixed 3,100 Mm^3. Alborzi et al. (2018: 10) argue that 're-establishing Lake Urmia's ecological integrity requires *aggressive* restoration policies and action plans aimed at maintaining environmental inflows'. Although at first glance this may appear ecocentric, the authors can be marked by partnership ethics, as they wish to develop an environmental water flow that is in accordance with human water needs. The researchers' estimate of when ecological equilibrium could be achieved varies from 3 to 16 years, depending on different scenarios (Alborzi et al. 2018: 9).

In contrast, Haghighi et al. (2018) include farmers' behaviour in their assessment of the environmental flow into Lake Urmia. Naturally, the agricultural sector depends on water from the river regimes that are supposed to help restore the lake. However, as farmers use as much water as possible during the irrigation season in May, the environmental flow would not suffice to meet the ecological water balance. Any attempt to regulate or legally prevent farmers or other water users from withdrawing river water would be 'difficult or [next] to impossible' (Haghighi et al., 2018: 3608). Environmental water should be released in the shortest possible time in winter and spring, specifically from November to May, when crop water demand is at its lowest. In addition, the researchers considered river ecosystems, which also need water to sustain themselves, as Lake Urmia cannot use environmental water exclusively for itself. As Haghighi et al. (2018: 3607) seem to advocate for both the hydrological environment, which includes the water needs of both Lake Urmia and rivers for a healthy ecosystem, and the agricultural sector, they can be placed in the category of partnership ethics. Haghighi et al. (2018: 3607) emphasise the need to develop environmental policies that also contribute to the preservation of 'agricultural land and jobs [...] in the Lake Urmia basin'. With this in mind, their assessment of the environmental water flow ranges from 2,400 to 3,400 Mm^3.

In a similar partnership-ethical approach, Moghaddasi et al. (2017) developed a drought warning system to assess environmental flows in a trade-off between Lake Urmia and the agricultural sector. This system combines the two aforementioned approaches by implementing both drought severity and agricultural water demand. Even in the worst drought scenario, water would be allocated to orchards but not to other annual crops and, more importantly, not to the lake. The prioritisation of orchards therefore highlights a slight preference for an anthropocentric position, which is encouraged by the fact that the researchers emphasise the drastic consequences for the agricultural sector: if current management strategies insist on a fixed annual environmental water flow 3,100 Mm^3 to reach an ecological water balance, then the agricultural sector is in distress. Assuming future hydrological scenarios that equate to the previous 50 years—that is ignoring increasing future effects of climate change—then the agricultural sector experiences drought 50% of the time. In other words, the environmental flow implemented by management strategies, with their static, ecocentric plans, would eventually bankrupt the agricultural sector (Moghaddasi et al., 2017: 449). Furthermore, the authors suggest that if current water management strategies are maintained, Lake Urmia will cease to exist within a decade, i.e., by around 2037. Restoration to a 'normal' level, even when positive changes are applied and climate change considered, would probably take more than two decades, i.e., circa 2057 (Moghaddasi et al., 2017: 449).

A report by the Japan International Cooperation Agency (JICA) calculated the necessary river inflow for Lake Urmia to reach a water level of 1,274.1 m. Their data showed that the lake requires a volume input of 4200 Mm^3 annually, which again is significantly more than suggested by the original assessment of 3,100 Mm^3. However, there are two issues here. First, according to the report, in 2020, the river inflow was approx. 1,900 Mm^3, which means that any water management strategy would need to increase river flows by 2,300 Mm^3. Second, even if this amount of water could be delivered to the lake, the JICA (2020: 8) suggests that it would take 30 years (2054) for restoration efforts to succeed. If

5,000 Mm^3 could somehow be accumulated in the lake, restoration may be achieved within a decade.

Finally, the ecological flow to meet the ecological water level was heavily contested even in the early days. In 2010, several meetings were held between national and provincial stakeholders (see Hashemi, 2012: 177), including the Ministry of Energy and Water, the Department of Environment, the Provincial Governor's Office (West Azerbaijan, East Azerbaijan, Kurdistan) and Agricultural Jihad Organisations. Although the ecological water requirement for Lake Urmia was established by law, the exact amount of water to be allocated by each of the three provinces had to be negotiated. However, even then, a water balance model by the MoE indicated that Lake Urmia would indeed need at least 3,900 Mm^3 per year to reach the ecological water level. According to Hashemi (2012: 193), who was involved in the decision-making process, the lack of transparency on the part of the MoE, which prevented insight into its model, and the divergent interests of the other stakeholders, meant that the 3,900 Mm^3 figure was neglected. The process unveils another interesting fact: despite the MoE's seemingly ecocentric, 'emotional appeal' to implement the 3,900 Mm^3 quota for Lake Urmia, and to limit water extraction for provinces in the basin to 2,900 Mm^3 per year, two observations were made by Hashemi (2012: 194). First, the predominantly anthropocentric position of most actors, and second, the covert anthropocentrism of the MoE. Thus, Hashemi (2012: 194) writes:

> 'The Commission made a staunch attempt to bring the "environment" into decision-making but the other actors had other priorities. WPAC's [MoE's representative] motive was not purely environmental: the Commission wanted to block extra applications by the provinces for more allocation and dam-building projects in the basin as well'.

However, the MoE's attempt was overruled and the 3,100 Mm^3 was agreed on. Overall, Hashemi (2012: 197) concludes that the LU water allocation process was controversial, as the whole process was 'complex and the manipulation of data and knowledge to suit negotiating positions caused a great deal of polarisation of opinions'. This highlights how politically manoeuvred the, at first, seemingly 'technical' and 'objective' approaches were. In the end, Lake Urmia received its water rights for 3,100 Mm^3, with West and East Azerbaijan allocated 2,035 Mm^3 and

1,080 Mm^3, respectively, and Kurdistan 585 Mm^3 despite being disadvantaged by the other decision-makers. Lake Urmia's share of water, however, lacked a sound and scientific assessment, leading Hashemi (2012) to conclude that fixed water allocations for any actor—including Lake Urmia—were not sustainable in the face of climate variability. Moreover, the total run-off that was estimated in the decision-making process at 6,900 Mm^3 was, in reality, according to Hashemi's calculations, 5,000 Mm^3, meaning that the amount of water allocated to each actor actually overshot the hydrological situation in the Lake Urmia basin (Hashemi, 2012: 222).

Although, currently, the ecological water level has normalised in the Lake Urmia discourse, it has always been a rather controversial issue. It has thus far proved impossible for the ULRP to achieve the designated water level until 2023, by a significant margin. In both ULRP reports (2015: 22; 2018: 74), environmental flow into Lake Urmia was planned to—in theory—successively increase from 1,800 Mm^3 in 2014/15 to 3,110 Mm^3 in 2018/19 and finally 4,380 Mm^3. In the final restoration in 2023/24, on average, the plan would allocate 3,300 Mm^3 per year, which again is less than the amount demanded by other studies.

On the ULRP homepage, Lake Urmia's water level stood at 1,270 m as of November 2022. For the aforementioned researchers, it has become clear that the water level as an indicator for the lake's health was a failure because it did not include other factors such as agriculture and climate variances and instead exclusively focused on a 'simple pyramid' of NaCl, *A. urmiana* and water level (or quantity), and because the set water level normalises a static objective that misses any flexibility in accordance with its dynamic human and non-human actors. Thus, the 'ULRP timeline [was] overambitious' (Alborzi 2018: 7), due to the simplifications made and the underestimation of the water requirement to reach the desired water level.

The scientific and political manoeuvring of Lake Urmia's water allocation that was determined and set by law can be considered—in terms of both theory and in practice—a failure in the view of most single water level sceptics, especially by Sima et al. (2021) and Parsinejad et al. (2022). The sheer variety of proposed environmental water requirements for Lake Urmia in recent scientific findings renders the set ecological water level as both a slight win, as Lake Urmia receives water rights, and yet a lose-lose situation because by default neither human nor non-human actors will receive the amount of water on which they rely.

5.3 The Many-Faced Causeway

Having examined the debate about the tolerance or vulnerability of Lake Urmia's aquatic ecosystem, and examined whether a single-target water level or multiple indices are better suited to restoring—or rather revitalising—the lake, there exists another, similar controversy, albeit with a different focal point. The water level debate revolved around water's potency to support life sustainably despite the unfavourable odds given in this scenario.

After an excursion from *A. urmiana* to the scientific debate, and then to the political meanderings of water rights and water volumes, I return to Lake Urmia again and to one of its controversially debated infrastructures. The Shahid (or Martyr, according to some literature) Kalantari causeway, which is half dike, half bridge, has sparked a debate about the vulnerability of Lake Urmia, but this time in terms of its vulnerability as a hydrological, abiotic 'object'. Put more simply, does the dam cause significant changes in Lake Urmia's hydrological system to such a degree that both the lake and its aquatic ecosystem are vulnerable to degradation, and ultimately extinction? Or is the lake's water capable of adapting to human intervention in a way that allows it to survive, and thus all the other actors directly linked to it (e.g., the brine shrimp, flora and fauna)?

The 'controversial causeway', as Madani (2014: 316) put it, was gradually constructed from 1979 to 1992 to cross the narrowest section of Lake Urmia from Sefi Gonbad port in the east on Islami Island to the western shore. The aim was to shorten the transport route between the cities Tabriz and Urmia by 130 km (Zeinoddini et al., 2009: 13; Zeinoddini et al., 2014: 4954). However, 30 years elapsed between the initial phase and the final construction before ferries were no longer needed to transport traffic across the remaining gap. From 2004 to 2008, a bridge structure was constructed and spanned a 1.25 km opening designed to maintain water circulation between the southern and northern parts of the lake (Zeinoddini et al., 2014: 4954). The dam consists mainly of earthworks, less than 15% steel and concrete piers, which are said to be harmful to the environment (Merufinia et al., 2014: 280). In the eyes of the public, the dam is mainly associated in a negative way with Lake Urmia.

In the media, the dam was mentioned in nine articles (out of 100 analysed) as one of the main causes of Lake Urmia's drying up, alongside other factors such as climate change, dam-building and poor water management (Havasi, 2018; Financial Tribune, 2017; Azernews, 2018; RadioFarda, 2018; Hosseini, 2022). The causeway literally 'cut' the Lake into two separate parts, thus interrupting the circulation of salt and fresh water between the two parts (Havasi, 2018; Hosseini, 2022), while the smaller parts were hence threatened by increasing evaporation (Azernews, 2018). Overall, the causeway threatens the ecological balance (Mehr News Agency, 2010) and is therefore stigmatised as a major antagonistic actor. In addition, an article published in 2012 offered a very different view of the causeway. Rajabova (2012)—working for Azeri.news—wrote that it 'has been suggested that [the causeway] has decreased circulation within the lake and altered the pattern of water chemistry; however, evidence suggests that the impact of the causeway on the uniformity of water chemistry in the lake has been minimal'. Apparently, the uncertainty Rajabova (2012) expressed must have vanished by 2017–2018, because afterwards the causeway was unquestionably treated as a major threat to Lake Urmia and the environment. Although the causeway has not received as much attention as other factors, its negative connotations persist, especially on the local level.

From a scientific perspective, Rouw (2017: 39) located the controversial causeway in the public debate primarily within what she called 'the natural frame', which is characterised by people's concerns that any human intervention in nature disturbs the natural order in a harmful manner. Farmers and NGOs are critical of these anthropogenic imprints on the landscape, or in this case waterscape, because natural processes are God-given and should be left untouched as much as possible, which hints at a religious rationale in this line of argumentation. Moreover, the causeway in particular violates one of the unwritten natural laws: the unity of nature. Interviewees were convinced that the causeway would violate this law and thus have a negative impact on the lake and the ecosystem. Two specific arguments were used to demonstrate this violation of the natural order of Lake Urmia. First, it would divide the lake into two separate parts, thereby disrupting the natural water circulation between them, and second, the springs under the lake would be blocked by the concrete, preventing the inflow of water (Rouw, 2017: 33). The latter reasoning, however, needs more explanation because a crucial water concept has been disclosed by studies conducting interviews with farmers. Pouladi et al. (2021) elucidated their hydraulic concepts of Lake Urmia, in that they think it is fed by springs underneath the lake, which are one of the primary sources of inflow. The causeway and the heavy equipment used during the construction caused the lake to shrink, as they 'created holes in the lake bed [sic], causing losses of lake water to underground caverns' (Pouladi et al., 2021: 8). As these interviews suggest, the farmers' idea of a water cycle includes a bottom-up process from beneath the lake, similar to the classical Greek water cycle proposed by Aristotle (Linton 2010: 112–113). This natural cycle, however, has been disturbed and even disrupted by the causeway and its construction. Similarly, the farmers interviewed for this thesis argued that in their opinion the causeway might be a root cause of Lake Urmia's desiccation. In addition to the farmers and NGOs, even local engineers expressed their concerns in relation to the causeway (Ranjbar, 2022: 6), moving the debate to the scientific platform which is far less unanimous about the impact of the causeway.

The aforementioned media perspective on the causeway reflects a somewhat similar pattern in the scientific community (Fig. 5.1). Of the 44 analysed papers, 26 presented the causeway as a negative or an even detrimental phenomenon, and 13 attributed either minimal, no effect or even a positive influence on Lake Urmia, while five had a neutral opinion. However, apart from the distribution of pro- and contra-causeway camps—or rather mostly not-contra or less-contra instead of pro—the chronological observation indicates three distinguishable periods in the causeway controversy. The first phase, from 2006 to 2013, shall be described as the 'non-contra' phase, which means that the causeway's effect on Lake Urmia was perceived as insignificant. The second phase (2014–2019) shows a sudden shift in favour of a contra-causeway position. Lastly and thirdly, a real scientific debate can be observed from 2020 to 2023, whereby studies argued against each other's findings based on empirical data and not based on opinions and quasi-facts. The causeway controversy, as will be demonstrated, proves the 'messi' nature of socio-technological processes and debates.

Prior to 2014, only Zarghami (2011: 90) had expressed concerns about the causeway, in that circulation and salinity patterns had changed to the detriment of Lake Urmia, its ecology and, first and foremost, *A. urmiana*. The literature he cited to prove his case was questionable, as all three scientists were indeed renowned *Artemia* experts but were not investigating the effects of the causeway on salinity patterns but on *A. urmiana*'s behaviour in relation to changing salinity concentrations (Abatzopoulos et al., 2006; Abbaspour & Nazaridoust, 2007). While Zarghami (2011: 9; emphasis added) assured that 'in early 2003, climatic conditions *and* the causeway caused the salinity of the lake to

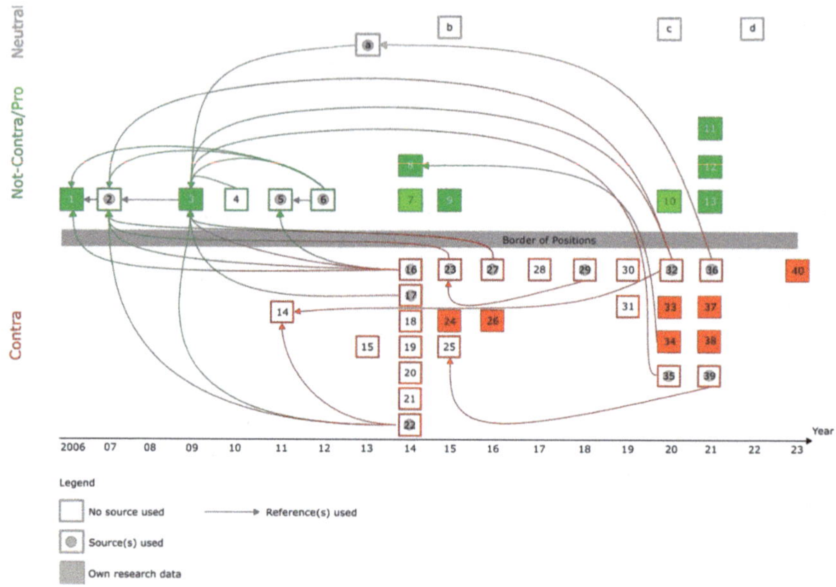

1 Alipour et al.
2 Eimanifar & Mohebbi
3 Zeinoddini et al.
4 Karbassi et al.
5 Golabian
6 Pengra et al.
7 Marjani & Jamali
8 Zeinoddini et al.
9 Zeinoddini et al.
10 Rosenberg et al.
11 Bayati & Danesh-Yazdi
12 Mojtahedi et al.
13 Dashzadeh et al.
14 Zarghami
15 Khatami & Berndtsson
16 Alipour & Olya
17 Zarrineh & Abad
18 Merufinia
19 Zoljoodi
20 Salehnia et al.
21 AghaKouchak et al.
22 Khalyani et al.
23 Tourian
24 Ghadimi & Nezammahalleh
25 ULRP
26 Hossieni & Ahmed
27 Alisadeh-Choobari et al.
28 Dalby & Moussavi
29 Mardi et al.
30 ULRP
31 Danesh-Y. & Ataie-A.
32 Ženko
33 Savafri et al.
34 Sabbagh-Yazdi et al.
35 Hosseini-Moghari et al.
36 Hemmati et al.
37 Bayati et al.
38 Barhagh et al.
39 Nikraftar
40 Mohammadi et al.
a Sima et al.
b Anvari & Valaie
c Saemian et al.
d Parsinejad et al.

Fig. 5.1 The misuse of quotations in scientific papers on the effect of the causeway

increase', Abatzopoulos et al. (2006: 301; emphasis added) expressed rather timidly 'the *possible* effects of this partitioning'. Nevertheless, Abatzopoulos et al. (2006) made a negative correlation between the causeway and *A. urmiana* that was possible, probable, but not empirically proven. Abbaspour and Nazaridoust (2007) did not even establish any connection between the causeway and *A. urmiana*. Overall, neither Zarghami (2011) nor his literature are sufficient to validate the causeway's real impact on Lake Urmia. In fact, by 2013, only a few studies had made the effort to analyse the causeway-lake relationship (see e.g., Sadra, 2004 in Parsinejad, 2022: 10), and as already mentioned, their view on the causeway was rather positive, or at least not negative. Conversely, there are some studies that see a positive impact of the causeway on Lake Urmia.

Eimanifar and Mohebbi (2007: 3), for instance, wrote in their review article on Lake Urmia how the causeway 'facilitate[s] water flow between the two parts' and that the opening gap is sufficient because 'there is a continuous water flow from south to north', due to strong currents. Therefore, the lake was deemed chemically uniform in both parts Alipour, 2006: 1). For the scientific community, however, the main peer-reviewed paper was 'Evaluation of dike-type causeway impacts on the flow and salinity regimes in Urmia Lake' by Zeinoddini et al. (2009), which approached the issue scientifically and not just based on common sense or anecdotal knowledge without reference. Their results suggest what the news article by Rajabova (2012) probably referred to:

'With respect to salinity patterns in Urmia Lake, it is concluded that over long term periods, the constructed *causeway has not had and would not have significant impacts on the overall salinity patterns* in Urmia Lake as compared to the natural (pre-causeway) conditions. Factors affecting the water balance of the lake (e.g., climatic variability, reduced river inputs) have had and will have the dominant role'. (Zeinoddini et al., 2009: 21; emphasis added).

A few studies picked up on this 'non-contra' stance towards the causeway because there was 'no evidence' (Golabian, 2011: 368) to prove the opposite (see e.g., UNEP & GEAS 2012: 129).

Since 2014, however, the tide has turned against the causeway; strangely so, in fact, and quite abruptly. This is odd in so far as no study challenged or scrutinised Zeinoddini et al.'s (2009) findings in 2014 or for the next six years—except for two studies. One of them, conducted by Ghadimi and Nezammahalleh (2015: 212), cautiously wrote that 'the causeway bridge can affect the lake hydrodynamic conditions'. More opposition came from a study by the Iraqi scientists Hossieni and Ahmed (2016), who had declared Lake Urmia's postponed death since the causeway was constructed in 2001, marking it the main cause of the catastrophe. However, both studies had a very low impact on the scientific community and therefore, I argue, did not catalyse the sudden anti-causeway movement.

Indeed, what did happen is as simple as it is disturbing. Numerous studies post-2013 had no evidence against the causeway but simply misquoted the aforementioned literature or even worse did not use any references at all and treated the causeway's negative impact as if it had been common knowledge all along. These allegations are serious, I admit, yet the following will prove them right. In this regard, I would like to make three reasonable arguments beforehand as to why this sudden shift might have occurred despite the lack of scientific validation. First, the political transition from the Ahmadinejad to the Rouhani administration took place precisely between 2013 and 2014, with the narrative shifting from a natural or climate-related disaster to an anthropogenic one and, more importantly, to a crisis exacerbated by human intervention within the basin rather than by foreign forces (see, for example, Madani, 2014, Dalby & Moussavi, 2016). Therefore, any man-made infrastructure that could affect the lake's environment was automatically pilloried. Second, scientists turned against the causeway because it was tempting to follow the simple logic that a causeway with an opening of 1.25 km must be doing something negative to the lake, and it was easier to blame something so seemingly and blatantly obvious than to acknowledge the fact that no consensus had been reached on the overall effect of the causeway. Against this backdrop, and third, the references may have been misquoted deliberately, but it is reasonable to assume that it could have been done inadvertently, because the articles cited were in some cases dubious and ambiguous in their wording and open to misinterpretation—and it is the latter with which I shall begin.

In 2014, nine 'contra-causeway' articles were published. According to these articles, the causeway was an 'irresponsible construction [that] made natural and fundamental changes in the hydrodynamic and ecological characteristics of the lake region' (Merufinia et al., 2014: 277) because it disturbed the normal water cycle (Zoljoodi & Didevarasl, 2014: 367). Therefore, the causeway 'aggravated the situation', marking it as an 'unsustainable activity' (Salehnia et al., 2014: 272) similar to other infrastructure resulting from 'aggressive hydro-economic development plans' (AghaKouchak et al., 2015: 309). The lake's desiccation and the causeway's inauguration in 2008 'simultaneously coincided' and intensified. Moreover, '[t]he flow and salinity regimes [were] affected by the presence of this new causeway' (Zarrineh & Abad, 2014: 43; citing Zeinoddini et al., 2009). Overall, Alipour and Olya (2014: 4) summarise that '[b]y reducing the circulation within the lake and altering the pattern of water chemistry, the causeway has accelerated evaporation'; meanwhile the authors cited *all* the articles stating the exact *opposite* (Alipour, 2006; Eimanifar & Mohebbi, 2007; Golabian, 2011; Zeinoddini et al., 2009). In subsequent years, this trend of not citing or misquoting validated science continued. For example, Tourian et al. (2015: 350) argued that the causeway caused the southern part of Lake Urmia to dry up, while Alizadeh-Choobari et al. (2016: n.p.) linked it to 60% higher salinity in the northern compared to the southern part of the lake. Similarly, Ženko (2020: 11) attributed a negative impact of the causeway on Lake Urmia, citing, among others, Eimanifar and Mohebbi (2007). All these misquotations and negative connotations relating to the causeway were even less reasonable, with another study on the 'Role of exchange flow in salt water balance of Urmia Lake' by Marjani and Jamali (2014: 14) concluding that 'the exchange flow through the opening *diminishes* the water level and salinity *differences* of the two lakes and hence helps restoration of the pre-construction state of the lake by the end of fall', thereby making it probably the only real 'pro' causeway study. This was followed by further studies by Zeinoddini et al. (2014, 2015) who reasserted their previous findings by stating 'The overall impacts of the causeway (by itself) on this hypersaline ecosystem and its unique shrimp living conditions are likely to be *much less significant than is commonly believed*' (Zeinoddini et al., 2014: 4969; emphasis added).

So how did it happen that the second phase of the dam controversy involved such a drastic shift and an apparently scientifically questionable reference standard? As mentioned above, it is probably a combination of all three factors: it was fashionable, seemingly logical and a simple mistake. Eimanifar and Mohebbi (2007) as well as Zeinoddini et al. (2009, 2014; 2015) are among the most cited and influential studies in the Lake Urmia controversy. In terms of their statements on the causeway, they can be misunderstood, if not read entirely and carefully. These papers may prove that even scientifically written inquiries turn out to be mediators rather than clearcut intermediaries; that is to say, they cannot—at least not in every case—convey knowledge without detours and transformations. Some recipients mistakenly turn knowledge upside-down when the format, structure or wording is ambiguous. In this case, Eimanifar and Mohebbi (2007: 3) may have been misquoted, because in the passage where they discuss the dam at Lake Urmia—and conclude that it has had minimal or no impact—they immediately turn to the Pacific Railroad, a causeway in a 'sister' lake, namely the Great Salt Lake in Utah, USA. There, evidence suggests that the causeway has increased salinity in the northern arm by 60% compared to the southern arm and has overall harmed the Great Salt Lake and its aquatic ecosystem. This second passage casts the causeway at the Great Salt Lake in a negative light. The mix-up between Lake Urmia and the Great Salt Lake was even mixed up by Zeinoddini et al. (2009: 21), who mistakenly wrote 'Salinity is approximately 60% higher in the northern basin of [Lake Urmia] because of higher evaporation rates and lower water river inflows than in the southern basin'. This might be nit-picking, and it yet shows how easily information can be mistakenly transformed. The other, often misquoted paper by Zeinoddini et al. (2009, 2014; 2015) may have caused ambiguity, as they appear paradoxical in terms of their abstract and introductory statements, as well as in their conclusion. For example, Zeinoddini et al. (2009) writes in the abstract that:

> 'The causeway [...] restricts water exchange. The flow and salinity regimes are affected by the presence of this new causeway, and there are concerns over the [sic] well being of the *Artemia* population'.

The abstract, however, fails to resolve this issue, as it does not mention the conclusion of the study.

In Zeinoddini et al. (2014), the paradoxical statements become even more disturbing. In the introduction, the causeway significantly influenced Lake Urmia's 'flow and salinity regimes [...] [while] [t]he circulation and mixing process of saline and fresh waters [...] [were] changed, and this change is generally *thought* to have resulted in higher salinities in the north and lower salinities in the south' (Zeinoddini et al. 2013: 79; emphasis added). 'Thought'—another word for 'not knowing'—was even substituted by the word 'believe' in a similar sentence in Zeinoddini et al. (2014: 4955) to clearly stigmatise the common misconception. As seen above, this false 'believe' unfortunately spread through the Lake Urmia scientific community, and so the causeway transformed from no impact

(1st phase) to, in the worst case, a significant factor in Lake Urmia's desiccation (2nd phase).

Although a study by the University of Tabriz addressed the effects of the causeway on Lake Urmia and salinity patterns in 2017—concluding a negative impact (see Parsinejad et al., 2022: 9)—scientific evidence on the matter of concern started to accumulate from 2020. However, despite growing interest in the causeway, there is still no consensus on whether it is as harmful to Lake Urmia and its aquatic ecosystem as is often propagated. What the scientists seem to agree on, though, is that it reduces water circulation between the two parts, and this has an impact on the chemical and biological properties of the lake, albeit they disagree on whether this impact is significant or minor (Parsinejad et al., 2022: 9). Hemmati et al. (2021), for example, argue that due to the causeway, salinity has increased in the northern arm, while Mohammadi et al. (2023: Abstract) further demonstrate that parts of the lake have 'almost independent hydro-chemical and hydrodynamic systems' and that this sort of anthropogenic intervention in 'sensible ecological systems' should have been better planned. Bayati & Danesh-Yazdi (2021), on the other hand, formulate a less drastic, but still negative impact of the causeway. Their study suggests a significant:

> '[I]mpact on the north–south water mixing when high flow rates enter the lake. This causes a one-month lag time between the occurrence of minimum [salinity concentration] in the southern and northern parts. However, the exchange of flow between the two parts is not remarkably influenced by the causeway in other seasons'. Bayati and Danesh-Yazdi (2021: 15).

Mojtahedi et al. (2022: 21) argue even more drastically in favour of the causeway, very much mirroring the findings of Zeinoddini et al. (2009, 2014), in that it has only a minimal 'impact on the uniformity of water in the lake', in which case their results 'are inconsistent with those obtained in previous studies, which have recognized the causeway as the leading cause of the crisis' (Mojtahedi et al. 2022: 21).

For Rosenberg et al. (2020), the causeway is not the antagonist of Lake Urmia's aquatic ecology but its saviour. Without it, they argue, the lake would be too saline, and thus *A. urmiana* would be incapable of surviving under these circumstances. This position differs substantially from other research, which therefore leaves a large question mark in terms of how to assess—especially as a social scientist—the 'real' effect of the causeway.

As a result of these divergent scientific findings, solutions for dealing with the causeway vary accordingly. On the one hand, the harshest critics suggest removing it altogether (Hossieni & Ahmed, 2016: 2) or, less aggressively, adding another opening to improve water circulation (Hemmati et al., 2021: 11) or creating a new opening in the middle of the causeway (Sabbagh-Yazdi, 2020: 1896)—which is currently closer to the eastern part. On the other hand, other findings suggest the exact opposite, namely that adding or even removing the dam would not stop the degradation of the lake (Zeinoddini et al., 2014: 4969; Dadashzadeh et al., 2021). The contradictory research results and suggestions help to conclude what Parsinejad et al. (2022: 10) already noted: 'More work is needed to identify the impact of the causeway on the lake ecosystem'.

The ULRP seemed to have recognised that the causeway was more of a controversial issue than a clear suspect in the degradation of Lake Urmia, as they were aware of the lack of knowledge on the issue (see ULRP, 2015: 16). Their plan for restoring the lake was to further study the causeway and offer solutions to its negative effects, with the Ministry of Environment taking the lead (ULRP, 2015: 34). Although the ULRP is aware of the causeway's impact on water circulation (ULRP, 2017: 11; ULRP, 2018: 67)—as is the scientific community—the continued lack of scientific consensus on the issue paralyses the organisation and forces it to do nothing, rather than take the socio-economic risk of acting.

5.4 Conclusions: A Shrimp as a Central Figure

The causeway and water level controversies have revealed a variety of conflicting water concepts. The most important observation, however, is that actors evaluate water concepts in relation to other non-human actors; i.e., the concern revolves not only around water, but also around the impact that other actors (are likely to) have on it. For example, Lake Urmia and its salinity concentration are synonymous with their ability to support life for the local brine shrimp A. urmiana, and therefore its entire aquatic ecosystem as well as the lake's surrounding ecosystem (e.g., wetlands, flamingos, mammals).

The debate on water salinity emphasises a certain spectrum of salinity concentration that allows A. urmiana to thrive. Hence scientists have decided on a minimum threshold at 1,274.1 m with a supposed salinity of 240 ppt NaCl, which corresponds to a salinity concentration of 263 ppt. At the other end of the spectrum, at a higher saline concentration, water loses its life-supporting potential and eventually turns into life-threatening water—even endangering humans, as we shall see. The techno-scientific process has exposed the political nature of the apparently objective number, i.e., the ecological water level, which may have been derived from scientific findings but is now—and has always been—contested by other scientists.

Critics of the single water level target point at the static assumptions used in the calculation, such as constant water availability, despite regular droughts in the region, and the increasing likelihood that the effects of climate change will become more severe in the near future. Although Lake Urmia was institutionalised as an entity in the form of the ULRP, with exclusive water rights established by national law, time has shown that the water level target was doomed to fail from the outset, as the ecocentric position represented by the scientists and the ULRP could neither be imposed on other stakeholders (such as provincial governors and ministries), who also demanded water for their own interests, nor could it ever be implemented in this static fashion. The single water level target also reveals another water concept: water is supposed to do it all. Save the lake, save the environment and meet all human needs. The quantity of water alone should work wonders. Alas, water alone is not sufficient as an indicator of the well-being of Lake Urmia and its collective, human and non-human association.

Moreover, Lake Urmia is perceived as being prone to vulnerability. The latter point has been particularly evident in the causeway controversy. The causeway has been stigmatised by local residents, the media and the majority of scientists as—in some cases—the primary cause for Lake Urmia's desiccation despite the fact that scientific evidence did not support this belief for many years. Indeed, the causeway does affect water circulation between the northern and southern parts. However, for some scientists, Lake Urmia is not as fragile or as vulnerable to every human intervention as often propagated. Interestingly, according to Rosenberg et al. (2020), it might even have a positive effect on the lake's ecology, as it reduces salt concentration, which in turn benefits local brine shrimp reproduction. The lake and the causeway, according to this point of view, have a positive relationship in the sense that the construction improves the ecology of the site, while this moves the causeway away from an antagonistic actor to one of a supporting cast member. The still ongoing scientific debate on the matter will certainly concretise the causeway's impact on the lake and the environment, but as long as findings remain conflicted and no consensus can be reached, Lake Urmia should not be hastily marked as fragile as it is often portrayed. Nevertheless, regardless of whether the scientific community has reached agreement, it would still be challenged by the pristine nature concept of the local people, who share a religiously motivated desire to leave Lake Urmia as it is—or to return it to the state it once had, without the drastic human intervention. This traditional religious concept of water is based on the belief that Lake Urmia should have remained a single entity and not been divided by the causeway, as this intervention violates God's natural processes. Similarly, the construction of the causeway and its embankment—according to this concept—destroyed the water caverns below the lake, which are one of its primary water sources. The contra-causeway and traditional religious water concepts essentially share the nature ephemeral perception, so any intervention by humans in nature is per se harmful. The 'pro'-causeway group, however, scrutinises this specific vulnerability of Lake Urmia and its environment.

In terms of the ULRP's project management, two observations can be made. First, in the case of the causeway, the ULRP prefers inaction rather than action, as it seems to be aware of the controversial nature of the construction, which does not currently allow for simple solutions. Second, the ULRP insists on the ecological water balance and measures its own success or failure by fluctuations in the water level—as does the media. This water level has not been reached yet, which is why the ULRP's achievements are often questioned by some actors, as discussed in Chapter 7.

References

Abatzopoulos, T. J., Baxevanis, A. D., Triantaphyllidis, G. V., et al. (2006). Quality evaluation of Artemia urmiana Günther (Urmia Lake, Iran) with special emphasis on its particular cyst characteristics (International Study on Artemia LXIX). *Aquaculture, 254*(1–4), 442–454.

Abbaspour, M., & Nazaridoust, A. (2007). Determination of environmental water requirements of Lake Urmia, Iran: An ecological approach. *International Journal of Environmental Studies, 64*(2), 161–169.

Agh, N., van Stappen, G., Bossier, P., et al. (2008). Effects of salinity on survival, growth, reproductive and life span characteristics of Artemia populations from Urmia Lake and neighboring lagoons. *Pakistan Journal of Biological Sciences PJBS, 11*(2), 164–172.

AghaKouchak, A., Norouzi, H., Madani, K., et al. (2015). Aral Sea syndrome desiccates Lake Urmia: Call for action. *Journal of Great Lakes Research, 41*(1), 307–311.

Alborzi, A., Mirchi, A., Moftakhari, H., et al. (2018). Climate-informed environmental inflows to revive a drying lake facing meteorological and anthropogenic droughts. *Environmental Research Letters, 13*(8), 1.

Alipour, S. (2006). Hydrogeochemistry of seasonal variation of Urmia Salt Lake, Iran. *Saline Systems, 2*, 9.

Alizadeh-Choobari, O., Ahmadi-Givi, F., Mirzaei, N., & Owlad, E. (2016). Climate change and anthropogenic impacts on the rapid shrinkage of Lake Urmia. *International Journal of Climatology, 36*(13), 4276–4286.

Azernews. (2018). *Too late to save Middle East's largest saltwater lake, Urmia?* https://www.azernews.az/region/126004.html. Accessed Augest 12, 2025.

Bayati, M. & Danesh-Yazdi, M. (2021). Mapping the spatiotemporal variability of salinity in the hypersaline Lake Urmia using Sentinel-2 and Landsat-8 imagery. *Journal of Hydrology 595*, 126032.

Castree, N. (2001). Socializing nature: Theory, practice, and politics. In N. Castree, & B. Braun (Eds.), *Social Nature: Theory, Practice, and Politics* (pp. 1–21).

Dadashzadeh, M., Mojtahedi, A., & Parsa, J. (2021). *Evaluation the Impacts of Causeway Remedial Actions on Urmia Lake Ecosystem Using Computational Fluid Dynamics and Spatial Analysis Method, 3*(51), 41–53.

Eimanifar, A., & Mohebbi, F. (2007). Urmia Lake (Northwest Iran): A brief review. *Saline Systems, 3*, 5.

Feitelson, E. (2017). A hierarchy of water needs and their implications for allocation mechanisms. In R. Ziegler & D. Groenfeldt (Eds.), *Global water ethics: towards a global ethics charter.* Routledge.

Financial Tribune. (2017). *Health Risks of Iran's Lake Urmia Desiccation Under Study*. https://financialtribune.com/articles/people/57489/health-risks-of-irans-lake-urmia-desiccation-under-study. Accessed Augest 3, 2024.

Ghadimi, M. & Nezammahalleh, M. A. (2015) Construction of a Causeway Bridge across the Lake Urmia and its Influence on Drying Trend of the Lake. *ISPRS - International Archives of the Photogrammetry, Remote Sensing and Spatial Information Sciences. XL-1-W5*, 211–213.

Golabian, H. (2011). Urumia Lake: Hydro-Ecological Stabilization and Permanence. In V. Badescu & R. B. Cathcart (Eds.), *Macro-engineering Seawater in Unique Environments* (pp. 365–397). Springer.

Groenfeldt, D. (2019). *Water Ethics: A Values Approach to Solving the Water Crisis* (2nd ed.). Routledge.

Haghighi, A. T., Fazel, N., Hekmatzadeh, A. A., & Klöve, B. (2018). Analysis of Effective Environmental Flow Release Strategies for Lake Urmia Restoration. *Water Resources Management, 32*(11), 3595–3609.

Hashemi, M. (2012). *A social-technical assessment framework for integrated water resources management (IRWM) in Lake Urmia Basin, Iran.* Doctor of Philosophy Thesis. Newcastle University.

Havasi, A. (2018) Iran sees 'revival' of imperilled Lake Urmia. *Phys.org,* 2018. https://phys.org/news/2018-12-iran-revival-imperilled-lake-urmia.html. Accessed Augest 3, 2024.

Hemmati, M., Ahmadi, H., Hamidi, S. A., & Naderkhanloo, V. (2021). Environmental effects of the causeway on water and salinity balance in Lake Urmia. *Regional Studies in Marine Science, 44,* 1.

Holman, K. (2021). *United States and Iranian Researchers Collaborate on Lake Urmia Restoration.* https://www.usu.edu/today/story/united-states-and-iranian-researchers-collaborate-on-lake-urmia-restoration. Accessed Augest 12, 2025.

Hosseini, M. (2022). *The Drying Up of Lake Urmia Will Destroy Millions of Lives and Hectares of Land - Iran Focus.* https://iranfocus.com/life-in-iran/49019-the-drying-up-of-lake-urmia-will-destroy-millions-of-lives-and-hectares-of-land/. Accessed Augest 3, 2024.

Hossieni, H., & Ahmed, S. S. (2016). Thermodynamics Reason for Dying of Urmia Lake, This is not Just an Aral Syndrome! *Journal of Geography & Natural Disasters, 06*(01), 1.

JICA. (2020). Data Collection Survey on the Improvement of the Hydrological Cycle Model in Urmia Lake Basin in the Islamic Republic of Iran: Final Report. https://openjicareport.jica.go.jp/pdf/12361952_01.pdf. Accessed Augest 14, 2024.

Kelts, K., & Sharabi, M. (1986). Holocene sedimentology of hypersaline Lake Urmia, Northwestern Iran. *Palaeogeography, Palaeoclimatology, Palaeoecology, 54*(1–4), 105–130.

Linton, J. (2010). *What is water?: The history of a modern abstraction.* UBC Press.

Madani, K. (2014). Water management in Iran: What is causing the looming crisis? *Journal of Environmental Studies and Sciences, 4*(4), 315–328.

Marjani, A., & Jamali, M. (2014). Role of exchange flow in salt water balance of Urmia Lake. *Dynamics of Atmospheres and Oceans, 65,* 1–16.

Mehr News Agency. (2010). Urmia Lake facing environmental crisis. *Mehr News Agency,* 2010. https://en.mehrnews.com/news/42779/Urmia-Lake-facing-environmental-crisis. Accessed Augest 3, 2024.

Merchant, C. (2017). *Science and nature: Past, present, and future.* Routledge.

Merufinia, E., Aram, A., & Esmaeili, F. (2014). Saving Lake Urmia: from Slogan to Reality (Challenges and Solutions). *Bulletin of Environment, Pharmacology and Life Sciences, 3*(3), 1.

Moghaddasi, M., Morid, S., Delavar, M., & Hossaini Safa, H. (2017). Lake Urmia Basin Drought Risk Management: A Trade-Off Between Environment and Agriculture. *Irrigation and Drainage, 66*(3), 439–450.

Mohammadi, A., Schwamborn, G. & Kaveh-Firouz, A. et al. (2023). Influence of dyke-type causeway on Urmia Lake (NW Iran); insights from water physico-chemical parameters seasonal (2019) changes. Journal of *Hydro-environment Research 47*, 1–14.

Mojtahedi, A., Dadashzadeh, M., Azizkhani, M., Mohammadian, A., & Almasi, R. (2022). Assessing climate and human activity effects on lake characteristics using spatio-temporal satellite data and an emotional neural network. *Environmental Earth Sciences, 81*(3), 1.

MoU. (2010). *Integrated Management Plan for Lake Urmia Basin*. https://www.ramsar.org/sites/default/files/documents/library/lakeurmiamanagementplan-i.r.iran2010.pdf. Accessed Augest 18, 2025.

Parsinejad, M., Rosenberg, D. E., Ghale, Y. A. G., et al. (2022). 40-years of Lake Urmia restoration research: Review, synthesis and next steps. *The Science of the Total Environment, 832*, 1.

Pouladi, P., Badiezadeh, S., Pouladi, M., et al. (2021). Interconnected governance and social barriers impeding the restoration process of Lake Urmia. *Journal of Hydrology, 598*, 1.

RadioFarda. (2018). Drought And Mismanagement Continue To Deplete Iran's Lake Urmia. *Iran News By Radio Farda,* 2018. https://en.radiofarda.com/a/29151393.html. Accessed Augest 3, 2024.

Rajabova, S. (2012). *UN assists Iran to prevent fallout from drying lake*. https://www.azernews.az/region/45570.html. Accessed Augest 3, 2024.

Ranjbar, A. M. (2022). The greening of human rights in Iran: Lake Orumiyeh, human rights, and environmental justice. *Environment and Planning e: Nature and Space, 3*(5), 251484862211081.

Rosenberg, D. E., Parsinejad, M., Sima, S., Wurtsbaugh, W. A., Null, S. & Sorooshian, A. (2020) How Can Researchers and Managers Better Collaborate on Lake Urmia Restoration? *Workshop primer* (pp. 1–6).

Rouw M. (2017). Socially navigating through the Urmia Lake debate: Master Thesis. https://edepot.wur.nl/421717. Accessed Augest 15, 2024.

Sabbagh-Yazdi, S. R., GhelichKhany, L., & Kalhor, K. (2020). Numerical investigation of the effects of causeway opening configurations on horizontal currents of Lake Urmia. *International Journal of Environmental Science and Technology, 17*(4), 1885–1898.

Sadra, (2004). *Hydrodynamic & Hydraulic and Environmental (Sediment & Salinity) Investigation (No. TEP81449YGP10002), Design & Construction Of the Oromieh Lake Causeway*. Sadra Company, Tehran, Iran.

Salehnia, M., Hayati, B., Ghahremanzadeh, M., & Molaei, M. (2014). Estimating the Value of Improvement in Lake Urmia's Environmental Situation Using Choice Experiment. *International Journal of Agricultural Management and Development, 5*(4), 271–282.

Sima, S., Rosenberg, D. E., Wurtsbaugh, W. A., Null, S. E., & Kettenring, K. M. (2021). Managing Lake Urmia, Iran for diverse restoration objectives: Moving beyond a uniform target lake level. *Journal of Hydrology: Regional Studies, 35*(9), 1.

Tourian, M. J., Elmi, O., Chen, Q., Devaraju, B., Roohi, S., & Sneeuw, N. (2015). A spaceborne multisensor approach to monitor the desiccation of Lake Urmia in Iran. *Remote Sensing of Environment, 156*, 349–360.

Tvedt, T. & Oestigaard, T. (2011). *A History of the Ideas of Water: Deconstructing Nature and Constructing Society*. I.B. Tauris.

ULRP. (2015). *Urmia Lake Restoration Program: Brief Report and Projects Outline.*

ULRP. (2017). Urmia Lake Restoration Program: A Brief Report.

ULRP. (2018). Urmia Lake: lessons and challenges.

United Nations Environmental Programme (UNEP) & Global Environmental Alert Service (GEAS). (2012). The Drying of Iran's Lake Urmia and its Environmental Consequences. *Environmental Development, 2*, 128–137.

Yekom. (n.d.). *About Us – Yekom at a Glance*. https://www.yekom.com/about-us/yekom-at-a-glance/. Accessed May 14, 2024.

Zarghami, M. (2011). Effective watershed management; Case study of Urmia Lake, Iran. *Lake and Reservoir Management, 27*(1), 87–94.

Zarrineh, N., & Abad, M. A. N. (2014). Integrated water resources management in Iran: Environmental, socio-economic and political review of drought in Lake Urmia. *International Journal of Water Resources and Environmental Engineering, 6*(1), 40–48.

Zeinoddini, M., Bakhtiari, A., & Ehteshami, M. (2015). Long-term impacts from damming and water level manipulation on flow and salinity regimes in Lake Urmia, Iran. *Water and Environment Journal, 29*(1), 71–87.

Zeinoddini, M., Tofighi, M. A., & Vafaee, F. (2009). Evaluation of dike-type causeway impacts on the flow and salinity regimes in Urmia Lake, Iran. *Journal of Great Lakes Research, 35*(1), 13–22.

Zeinoddini, M., Tofighi, M. A., & Bakhtiari, A. (2014). Assessment of 2DH and pseudo-3D modelling platforms in a large saline aquatic system: Lake Urmia, Iran. *Hydrological Processes, 28*(18), 4953–4970.

Ženko, M. (2020). *Hydro-social transformations in the Lake Urmia basin, Iran*. https://ddd.uab.cat/pub/tesis/2021/hdl_10803_670989/mazel1del.pdf. Accessed September 21, 2024.

Zoljoodi, M., & Didevarasl, A. (2014). Water-Level Fluctuations of Urmia Lake: Relationship with the Long-Term Changes of Meteorological Variables (Solutions for Water-Crisis Management in Urmia Lake Basin). *Atmospheric and Climate Sciences, 04*(03), 358–368.

CHAPTER 6

The Prima Causa of Lake Urmia's Desiccation

Now that the ecological water level and causeway controversy have been unveiled in detail (Chapter 5), it is time to turn to the meta- or super-controversy (Venturini, 2012: 11) in which the causeway is embedded, namely the vexing question as to whether Lake Urmia's desiccation is primarily climate induced or due to anthropogenic factors. However, as both are anthropogenic in origin, as in anthropogenic climate change, I will continue to refer to them as endogenous (for the latter) and exogenous (for the former). The answer to this question is of utmost importance, as the leading narrative in this discourse—or controversy—dictates and justifies any measures taken to cope with Lake Urmia's critical state. From the government to the ULRP, scientists, the media and the local population, all of them diverge in their perspective on the leading causes of Lake Urmia's desiccation. Due to these diametrical oppositions, solutions that are presented by some actors are not well received by other actors, which in turn means that the controversy remains nothing but a heated debate with no resolution between the involved parties in sight.

Once more, the best place to follow the investigation is through the meta-analysis, the panorama, of Parsinejad et al. (2022), because this study attempts to bring a halt to the debate and finally end it for good. The issue, however, as will be shown, is that the causality established to reason in favour of their argument is not without issues. To find a better

R. Gonda, *Lake Urmia's Hydro-Social Dilemma*,
https://doi.org/10.1007/978-3-032-08941-0_6

footing in this controversy, a quick review of the media will help identify some narratives surrounding Lake Urmia's desiccation.

6.1 A Short-Circuited Debate of Humans vs. Nature: Panorama's View

Climate change, droughts, dams, the causeway, agriculture, groundwater depletion, inefficient irrigation, mismanagement, the government, the trade embargo, urbanisation and population growth are among the most renowned factors having an impact on Lake Urmia. Almost everyone agrees that each of these factors plays a role. The question remaining is, which one of these contributes primarily? The overall impression made by the 60 analysed media articles is that anthropogenic factors are blamed on a more regular basis than natural, exogenous factors. Overall, dams and agriculture were accused primarily, while droughts, mismanagement, climate change and the causeway followed in third, fourth and fifth place, respectively (Fig. 6.1). Natural factors, however, over time, decreased and were mentioned less in comparison to anthropogenic ones.

In the sciences (Fig. 6.2), the picture differs to some degree. Agriculture is by some distance the main issue for Lake Urmia's desiccation, and it has been identified, or accused as such, almost from the very beginning of scientific inquiry. The other factors that are regularly mentioned are dam construction, droughts, climate change and the causeway. These factors, similar to agriculture, have been mentioned on a regular basis. Moreover, they could be marked as primary causes, whereas the others could be marked as secondary; however, this assessment is nothing but quantitative and similar to the media analysis, in that it does not provide an answer to which one of these factors has caused the most trouble for Lake Urmia. Is it due to anthropogenic, local or natural—in the sense of exogenous—factors? This question remains unanswered.

Parsinejad et al. (2022: 4; emphasis added) conclude in their synthesis report that 'anthropogenic factors contributed to Lake Urmia level decline, while climate factors had a smaller or *no impact*'. The conclusion is definitive and the discourse is brought to a close, primarily as a result of scientific inconsistency.

One hundred and ten articles were analysed and subsumed to offer a final consensus on what two years earlier was still marked by Schulz et al. (2020) as a 'controversial debate' on the main cause. In order for Parsinejad et al. (2022) to convince readers of their findings, two

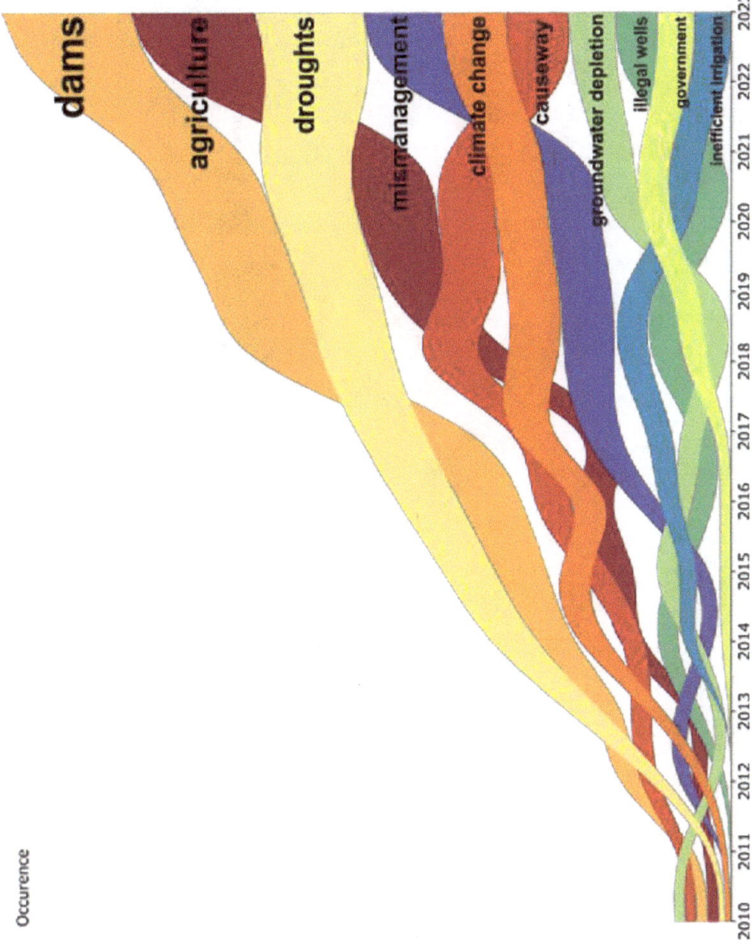

Fig. 6.1 Bumpchart (cumulative): causes of Lake Urmia's desiccation in the media ($n = 60$)

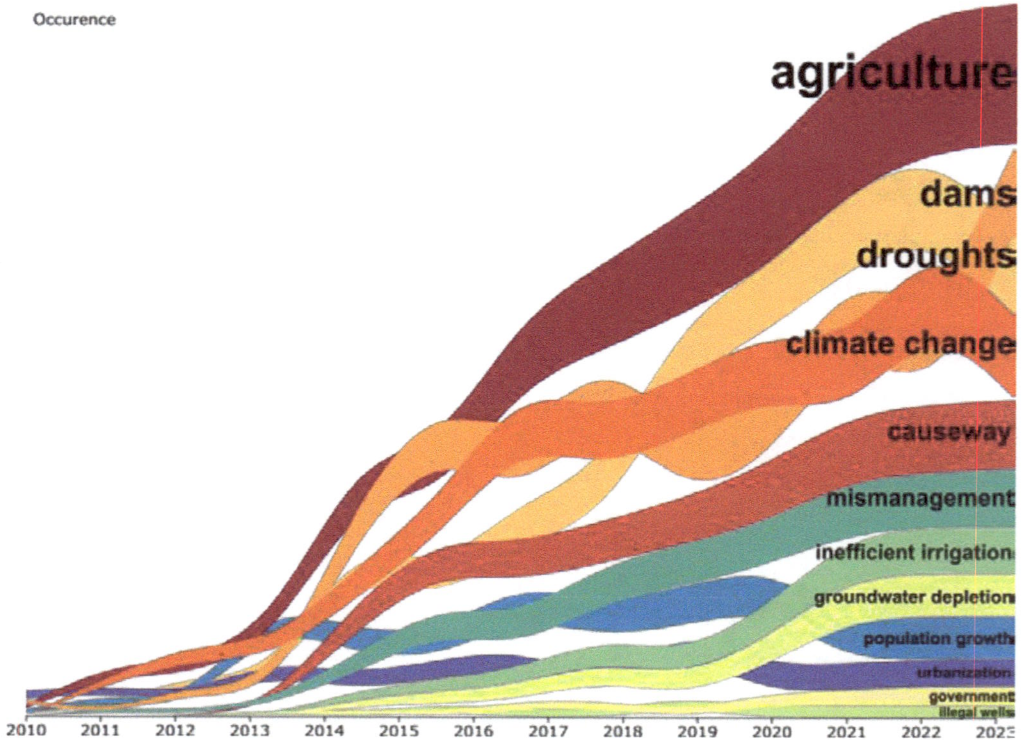

Fig. 6.2 Bumpchart (cumulative): causes of Lake Urmia's desiccation in the sciences ($n = 46$)

strategies were chosen: let the facts speak for themselves, or, rather, the naked numbers, and minimise the effect of climate change or natural factors as much as possible. Indeed, Parsinejad et al. (2022: 5) provide an overview of the percentage distribution between climate change and human impacts. The impression made in this regard clearly indicates, except for a few outliers, that anthropogenic factors drive Lake Urmia's desiccation and that this can be considered a general consensus in the sciences. Moreover, based on the second move, climate change is presented as having a minimum-to-non-existential impact on Lake Urmia. This argument, however, is outlined in a scientifically questionable fashion. A review and synthesis report surely is meant to gain as much of an 'overview' as possible, and identify a common thread across the scientific landscape. Alas, though, there is a significant difference between identifying similarities and *constructing* them. The politics of science become self-evident if we move closer to the argument made by Parsinejad et al. (2022) and dissect how they wish to convince readers that Lake Urmia is almost *exclusively* impacted by anthropogenic factors.

The first ill-constructed move is the argument proposed that although other scientists conclude that precipitation 'has contributed to the lake's decline [...] [our] analysis of precipitation shows there has been no significant decrease' (Parsinejad et al., 2022: 5). The scientific proof underlining this statement as a given fact comes with reference to a figure showing the relationship, or non-relationship, between Lake Urmia's water level, irrigated cropland and precipitation. The figure and its subtext conclude that the expansion of irrigated agricultural area has a direct correlation with Lake Urmia's decline, whereas precipitation has remained steady from 1985 to 2020 and therefore is excluded as a negative impact factor. To accentuate this position, in the appendix the mean annual basin precipitation is shown; however, there is no way to prove or to reconstruct this graphic. I am not questioning that this observation or conclusion might be incorrect, but the method as it is presented loses its scientific validity and drifts into the world of fiction because it lacks a chain of references and acts as if the illustration can simply speak for itself, without referencing the source material. Admittedly, placing the word 'fiction' next to 'science' could be perceived as the biggest insult any researcher could receive, but as Latour (2013: 251) has explained, for example, in his book 'Modes of Existence', these two *entities* have been entwined since the seventeenth century. To elucidate, researchers and their recipients alike require the assistance of fictional entities to conceptualise the actions occurring within the delineated milieu. As Latour (2013: 251) posited, such entities serve to 'transport' the imagination. The main difference between fiction and the crossing of fiction with science lies in the chain of reference that is established in the sciences to distance itself from a figurative world and establish traceable links (references) to a literal world. Latour (2013: 251–252) argues in his own words:

> 'If we authorize beings of fiction to travel far and wide [...] into another world, the same domesticated beings, disciplined by chains of reference, have to come back home to bring back, 'report on', remote states of affairs, which they also are responsible for unifying in a verifiable common narrative'.

Later on, he adds:

> 'The literal is nothing but the disciplined and domesticated replaying of the flight of figures. The literal is to the figurative what the dog in the fable is to the wolf. The sciences may be fictions, but they are domesticated enough to report, refer, inform'.

As made evident, science and fiction—or references and fiction in Latourian terms—act congruently; otherwise, how shall researchers report any of their findings to their recipients if neither of them can imagine their reports without the fictional entities? Yet, the difference between an exclusively fictional narrative and the 'domesticated' beings of fiction in their crossing with references lies in the fact that the latter have to go through a series of tests indicated by numerous references that assure us of their validity and connection to the literal world. Indeed, the controversy on the ecological water level, and even more so the causeway controversy, provide empirical proof relating to how 'scientific' findings lose their connection to the literal world if they do not establish a chain of references. In this sense, the causeway moved all the way from being non-significant to significant for Lake Urmia's ecosystem in a non-scientific manner characterised by misquoting and the assumption of a premature conclusion on the nature of the causeway. In fact, I could have made the little excursion to Latour's observation on fiction and references in the sciences in the previous chapter, but doing so now, I argue, clearly elucidates the political character of the sciences at Lake Urmia. As Parsinejad et al. (2022: 4), and others, understand:

'Determining anthropogenic and climate effects on the Lake Urmia basin is technically challenging and politically fraught because answering the question has the potential to blame one or more groups of people for lake decline'.

Returning to the statement, namely that precipitation has had no impact on Lake Urmia's decline, therefore has political implications: the less non-human actors are pinpointed as culprits, the more attention shifts towards human actors, and so—drastically speaking—they become enemy number one of the current hydro-social relation in a political sense and must be excluded or substituted by other, less harmful entities by any means. In other words, to describe simply *what is there* always has a moral implication, so the most objective, dispassionate observation turns into a moral lesson, a passionate prescription on *what ought to be* (Latour 2018: 22–23). The issue at this point, in this meta-controversy that revolves around the main cause or party (human or natural/ internal or external/ non-human or human) responsible, lies in the often non-objectivity nurturing the normative implications of which moral claims are derived. Parsinejad et al.'s (2022) article in this sense holds a powerful position due to its responsibility, which tries to convey that on the one hand it presents the scientific landscape of Lake Urmia—the panorama—as well as their findings as truthfully as possible, while on the other hand its synthesis always has moral and political implications governed by the full array of objective science under its belt. For this specific review article, as can be observed, however, it was not enough to simply state the numerous, sometimes diverging scientific findings but to *add* their *own findings* as well to fully underline their synthesis. This additional perspective in the causes controversy could have been valuable, but it has its flaws.

To grasp fully the extent of the narrative that is being pushed towards anthropogenic factors as the main issue at Lake Urmia, the overview of literature that was provided needs a second look. As explained above, the table provided by Parsinejad et al. (2022) is meant to speak for itself and leave any doubt right at the front door—again, numbers cannot lie. The latter is true in the sense that while numbers are just that, numbers, their arrangement and use do disclose a lot about the scientist using them. Take, for example, the very first paper published on the matter by Hassanzadeh et al. (2012: 129), whose conclusion proliferated all the way from the sciences, media and probably even to the tweet of Leonardo di Caprio in 2016, whereby Lake Urmia had become desiccated due to:

> '[...] climate change and overuse of surface water resources [as] the main factor [responsible] for 65% of the effect, constructing four dams [attributed to about 25%] of the problem, and less precipitation on lake has [an influence of an estimated] 10%'.

Attention should also be paid to the exact wording of the authors, whereby the reduced inflow into the lake is correlated primarily with the anthropogenic effect of 'overuse of surface water resources' *and* the impact of climate change. Both factors, natural and human, are implicit in the 65% attributed to the decline of Lake Urmia, albeit it cannot be numerically distinguished which one of them contributes more to this figure. In this regard, Parsinejad et al. (2022: 5) summarise that Hassanzadeh et al. (2012) arguably attribute 90% of the crisis to human impacts and only 10% to climate change—the 10% obviously relating to the decrease in precipitation. It is needless to say how distorted these percentages are in favour of the argument that anthropogenic factors outweigh climatic factors and as such emphasise a home-made, endogenous disaster. The question remains, though: Does this trend continue with other 'objective' numbers? Briefly, no. Yet, another issue arises.

Nine studies with numeric data were enlisted by Parsinejad et al. (2022: 5) on the main cause topic, including Hassanzadeh et al. (2012). Seven of them indicate that anthropogenic factors have a more significant overall impact on Lake Urmia's desiccation. These numbers were checked by the author, and two studies were misquoted again, and yet they should be regarded as unintentional mistakes. Hosseini-Moghari et al. (2020: 1954) did not view anthropogenic impacts on Lake Urmia as importantly as climate change:

> 'About 52 %–57 % of the total water storage loss in Lake Urmia basin and only 39 %–43 % of lake water loss during this time period was due to human water use, and 43 %– 48 % and 57 %–61 %, respectively, was due to climate variations'.

The study differentiated between water loss in the entire Lake Urmia basin and Lake Urmia, where human impact was more significant in the case of the former but less so for the latter. The second misquotation occurred with Kanani et al. (2020), this time, however, in favour of climate change, as the case study of the Lighvan sub-basin showed anthropogenic impacts (65–84%) as the main contributor to water loss.

6.2 Re-assessment of the Debate: An Unsettled Controversy

As the referencing mistakes of the scientific literature on the prima causa of Lake Urmia's desiccation occurred on a regular basis and, more importantly, Parsinejad et al. (2022) were clearly in favour of their ready-made argument that pro-anthropogenic factors were the leading cause of the disaster, an alternative 'overview' of the controversy shall be presented. Another main issue that cannot be left unmentioned is the fact that not every study that gained insights into the main cause published percentages but instead offered qualitative statements favouring climate (exogenous) or human (endogenous) impacts. Moreover, what is also of interest for this inquiry is identifying differences rather than consensus on this controversial topic; that is to say, how many studies not based on their own empirical research on the topic stated that anthropogenic impacts, climatic factors or both equally contributed to Lake Urmia's crisis. For this purpose, Fig. 6.3 was developed. The goal of this illustration is to categorise different papers by their positionality in regard to the main cause controversy. Studies that did provide percentages on the matter are highlighted by a fully coloured dot. These dots should be read alongside the secondary y-axis (right side) that shows the percentage ratio between climate (or exogenous) and anthropogenic (or endogenous) factors. Those studies that did provide qualitative statements are represented by a smaller dot within an empty, larger dot. In this case, studies of this type should be read alongside the primary y-axis (left side), whereby the analysed papers were distributed in a field indicating their positionality in certain degrees. The last category, covered studies that did not conduct research on the matter but took a position on what in their opinion, or to the best of their knowledge, drove Lake Urmia's desiccation. In the case of these studies, which did not provide percentages, statements were analysed, coded and categorised based on their wording. On this basis, they were placed in one of the following categories: equal, primary, exclusively. To give a few examples of how this system was applied, different statements will be presented alongside their chosen categorisation.

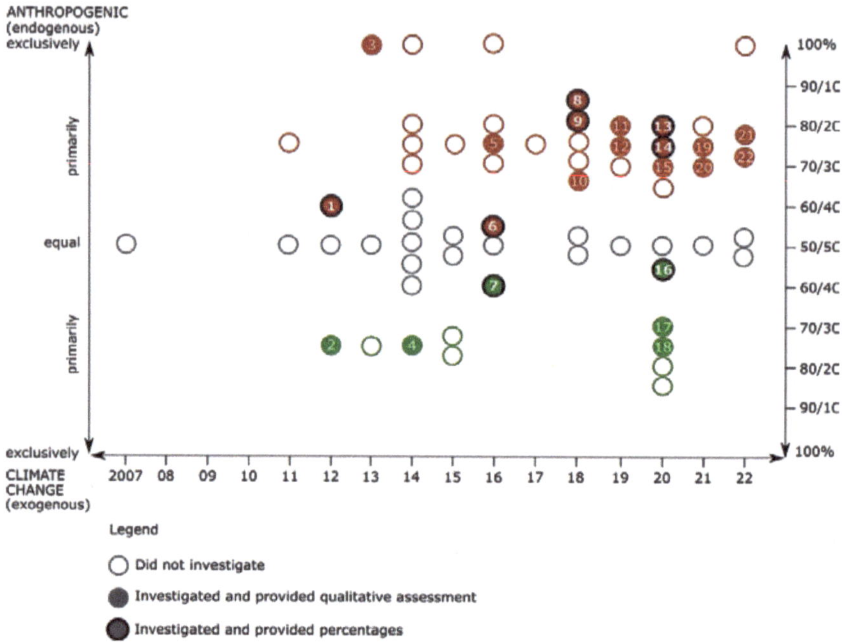

1	Hassanzadeh et al.	9	Alizade G. G. et al.	17	Schulz et al.
2	Abbaspour et al.	10	Alborzi et al.	18	Nhu et al.
3	Norouzi et al.	11	Khazaei et al.	19	Fouroumandi et al.
4	Fathian et al.	12	Dariane et al.	20	Rahimi & Breuste
5	Alizadeh-Choobari et al.	13	Pooralihossein & Delavar	21	Khoshnood et al.
6	Hamzekhani et al.	14	Zarghami	22	Mojtahedi et al.
7	Shadkam et al.	15	Sheibani et al.		
8	Chaudhari et al.	16	Hosseini-Moghari et al.		

Fig. 6.3 Overview of positions and research conducted on the prima causa of Lake Urmia's desiccation

However, before I delve into the excerpts and their supposedly fitting categories, an important remark must be recalled. First, I do not claim to know what the scientists really think of the disaster's main cause, as such I do not evaluate the researchers' opinion on this controversy, but their papers'; as Latour (2004: 226) wrote: 'The actors do not know what they are doing, still less the sociologists. What manipulates the actors is unknown to everyone, including researchers in the social sciences'. The scientists may quite possibly disagree with the positionality I attribute to them in Fig. 6.3 and the following examples and rightfully so. Yet, the effort was made in the spirit Venturini's (2010: 5) suggestion that 'you shall simplify complexity respectfully'. The simplification, however, is not aimed at the researchers themselves, but their papers because, secondly, texts are perceived as actors, hence every text has an agency (Nimmo, 2011: 114), a way to represent—rather than present—the common world (Collins & Pinch, 1993: 147) or the Lake Urmia case in this context. The scientists may differ in their personal view from my estimation, in terms of whether they attribute anthropogenic climate change or the anthropogenic impact in situ as the main driving factor.

Nonetheless, texts, sometimes even plainly, speak for themselves and reveal whether the author intended to do so or not. In other words, texts develop a self-sufficient agency independently from researchers, and hence their opinions on the matter are manoeuvred—possibly, not with certainty—into one of the aforementioned categories. The differentiation between researchers and their proposed research findings presented in the papers must henceforth be recalled regularly by the recipients. Otherwise, one may commit the fallacy of pretending to know what researchers think only by pinpointing their text, arguing: 'Your text resembles what you think; it mirrors your mind'. This fallacy is what I wish to avoid to the best of my abilities. In order to provide as much transparency as possible for the scientific landscape in the causes controversy, statements—if their chosen category seems ambiguous and doubtful—will be elucidated in the context of the whole text. Against this backdrop, the following categories (equal, primarily, exclusively) were applied to 41 papers, and based on these findings, Fig. 6.3 was developed.

The first category presented is 'equal'. If a study enlisted anthropogenic and natural factors but did not lean to one side or the other, then this category was applied as a code. For example, Azarnivand et al. (2015: 13) wrote:

> 'Lake Urmia in Iran is plagued by natural and anthropogenic driving forces'.

In similar fashion, Zarghami and AmirRahimi (2017: 310) acknowledged:

> 'The main reasons for this decline in water levels are still being challenged by academics and researchers'.

These two papers were located as 'equally positioned' in the controversial debate, because the former statement equalised both drivers' impacts, while the latter confirmed the heated debate, which has not reached consensus yet—at least in 2017. The next excerpt proves slightly challenging:

> 'Exacerbated by long drought periods, LU has experienced a long sever condition for more than 20 years due to the intense disruptive anthropogenic activities and mismanagement of water resources in LUB'. (Danesh-Yazdi & Ataie-Ashtiani, 2019: 650)

The statement certainly emphasises the anthropogenic impact but simultaneously mentions climatic change as an equally important stressor, without leaning to any side in particular. Put differently, since both factors were elevated and equally impactful on Lake Urmia, the paper was categorised as 'equal'.

Consequently, if a study argued that Lake Urmia's desiccation was mainly due to climate change or anthropogenic impacts, it was categorised as 'primarily'. It is important to note the vagueness of such a category, as it shifts the controversy into a rather polarised landscape. Regardless, it implies to which one of the two factors they tend to lean more, without excluding the other entirely. See, for example, a statement made by Tourian et al. (2015: 358):

> 'We conclude, from our results, that the drying up of the lake has been occurring due to a chain of reasons, which are highly influenced by anthropogenic activities'.

Here, anthropogenic factors are designated a more significant impact than natural factors, such as climate change or reoccurring droughts. Yet, due to the formulation 'highly influenced', and the fact that the authors did mention natural factors in their introduction, they cannot be marked as entirely excluded but perceived instead as a second-degree cause. Similarly, Ženko and Menga (2019) were categorised as 'primarily', although it is almost on the borderline in relation to a more extreme position. The paper may not exclude climate change as a driving factor, but it is reduced mostly to a narrative used in the discourse by the government to deflect from a manufactured drought. Hence, Ženko and Menga (2019: 12) classify Lake Urmia's desiccation as follows:

> 'It is interesting to note that the discourse of climate change has been used by the authorities both to justify the investments into hydraulic infrastructure as an adaptation policy to climate change, as well as to explain the cause for the degradation of the downstream areas, which is *in fact* primarily the result of poor water management decisions'.

In this case, the signal word 'primarily' marks this paper as the eponymous category suggests. To justify their statement, studies from Khazaei et al. (2019) and Madani (2014) were presented as a sufficient proof. Note how the paper tries to short-circuit the debate, which, as we have seen, has been characterised by others as controversial through the decisive use of the phrase 'in fact'. The study by Khazaei et al. (2019: 27) did not consider changes in precipitation and temperature as being 'primary causes of the lake drying so far'. Madani, who was also a co-author in Khazaei et al. (2019), is an important sciento-political figure, whose impact in Ženko and Menga (2019) is clearly evident in regard to the literature used to validate their arguments. We will turn our attention to Madani later on, but what should be noted for now is that Madani (2014: 323) did not provide debate-silencing 'facts' on the matter while commenting on Iran's general water crisis, whereby Lake Urmia was used as a prominent example of mismanagement and political failure. Ženko and Menga (2019) indicate that based on these references climate change and droughts may contribute to Lake Urmia's desiccation—which they do not mention directly but in-between lines—yet they minimise its effect relative to anthropogenic stressors and somewhat degrade climate change and droughts as a political tool used by water companies and the government.

The controversial nature of the causes of Lake Urmia's desiccation will be self-evident with the next and, more importantly, contradictory statement on the matter.

Schulz et al. (2020: 7) summarised their findings as follows:

> 'We could show that in the last decades variation in the volume of Lake Urmia was mainly triggered by changes in climatic conditions, and even without agricultural water extraction the general trend of the lake volume variations would have been the same. However, this conclusion does not mean that human influence on the hydrologic system in the Lake Urmia catchment is negligible'.

Although anthropogenic stressors are acknowledged, their impact is reduced to a second-degree cause, as climatic variations were identified as the main reason. The statement by Schulz et al. (2020) is almost antithetical to, e.g., Ženko and Menga (2019) and proves multiple, yet objective viewpoints on the same matter of concern, which is a characteristic of 'second-degree objectivity' (see Venturini, 2012: 798).

Another example of a 'primarily natural' position was the article by Nhu et al. (2020). Based on Landsat data, the article contradicted the argument that precipitation and temperature did not have a significant impact on Lake Urmia's desiccation. Their findings suggest an increase in temperature from 1986 to 2017 by 19% and a decrease in precipitation in the same period by 62% (Nhu et al., 2020). Now, the article would in fact be predisposed as neutrally positioned, as it does cite anthropogenic and natural stressors equally in the abstract; however, to prove their own findings, they turn to another supposedly similar study claiming that:

> '[...] the findings of this research that explain the reasons for shrinking of Lake Urmia are supported by Hassanzadeh et al. [(2012)] who demonstrated that climate change (65%), damming (25%), and low precipitation (10%) have contributed to the lake shrinking. In our study, we have also demonstrated that climate change and dam construction play significant roles'. (Nhu et al., 2020: 11)

Hassanzadeh et al. (2012), as one of the most prominent studies in research on Lake Urmia, really do experience some transformations—or unscientific treatment, one might say. As a reminder, the 65% figure was not exclusively attributed to climate change but also to 'overuse of surface water resources', as mentioned above, which clearly indicates human impact and as such an unknown yet relevant share of the 65%. Nhu et al.'s (2020) argument is hence wrong in the sense that their findings are backed by another study.

The overall treatment of scientific articles helps in making the following comment. The manipulation of data, whether intentional or not, could be polemically titled 'fictional-science politics': The use of references in this context represents a violation of the scientific method, whereby one is required to accurately represent the words of others and to cite them correctly. This is particularly concerning when one's conclusions are based on misquoted or paraphrased references, as this practice is tantamount to misrepresenting the scientific evidence. It is also concerning when a researcher's own findings are not validated, as this raises questions about the objectivity and rigor of their work. Finally, the use of selective quotations or data manipulation to fit a particular argument is a form of scientific fraud that is unacceptable in any field of research. The probably objective findings are on these few occasions embedded in literally constructed truths to imply that one's findings mean that one is walking on the right path. This does not suggest, though, that the specific findings of these articles are invalid but that their narrative is fictional and highly political. The tendency to use rather political claims compared to scientific ones will be self-evident in the next category.

The category 'exclusively' indicates that only one of the two was mentioned as Lake Urmia's stressor. The disregard of the other implies an extreme position that either leaves out exogenous/endogenous factors as a contributing cause or denies its effect by actively arguing against it. To give an example, for an active discredit, Chaharborj (2014: 302) argues:

> 'Although, in opinions of some experts and officials one of the main reasons for the drying up of Urmia Lake is drought and rising global temperatures, the UNEP also admits in its report that at distances less than 200 km from Lake Urmia, Lake Van is located in Turkey with a similar condition to Lake Urmia which continues to exist'.

Before Chaharborj (2014) argued against climate change or drought as possible stressors, anthropogenic factors were mentioned and as such indicated as the sole or exclusive reason for Lake Urmia's desiccation. However, without giving any context, the argument may sound invalid and questionable, as the category 'primarily' may also apply because it was at least mentioned from 'some experts' and officials' perspectives. The paper was categorised as 'exclusively anthropogenic impact' due to the neglect of climatic stressors caused by climate change or droughts, whereas dams, the causeway, mismanagement and an outdated agricultural system were blamed at the outset of the article, or rather the abstract. Moreover, in a second move, the paper tries to convince the recipient that drought and climate change are *not* by any means the cause of Lake Urmia's desiccation, as even the UNEP's report seemingly, or indirectly, delivers evidence to support this argument. This is achieved by making reference to the UNEP's comparison between Lake Van in Turkey (the UNEP & GEAS, 2012: 135) report also made mention of Lake Sevan in Armenia) and Lake Urmia. The latter appears to have responded in a sensitive manner to the dam policy in the basin, undergoing a significant transformation while the other lakes have remained relatively unchanged, at least according to satellite imagery comparing 2001 and 2011. However, the UNEP report also summarised that both anthropogenic and climatic factors caused Lake Urmia's desiccation (therefore, it was categorised as 'equal'). The paper by Chaharborj (2014), in other words, drives its agency clearly towards an exclusively anthropogenic position despite the diverging position of its cited reference.

In other cases, if climate change or drought were not mentioned at all, and therefore their impact was passively denied, they were similarly categorised as 'exclusively'. To give an example, Nouri et al. (2017: 1) note:

> 'Various factors explain the drying up of Urmia Lake. Demographic developments, human activities and especially demand for water in farming lands near the lake are among the main factors causing the lake's shrinkage'.

Although this sentence would suggest applying 'primarily' anthropogenic as the most accurate category, the authors failed completely to mention climate change or droughts as contributing drivers of the crisis. Recipients are therefore encouraged to perceive Lake Urmia exclusively as a human-induced disaster; hence 'exclusively' was applied here as the correct category.

Applying these categories to 62 scientific papers, Fig. 6.3 highlights the following observations. First, the majority of papers suggest that the main cause of Lake Urmia's desiccation is anthropogenic, while a minority point primarily at climatic factors. However, note that of the 16 studies siding with anthropogenic factors, six provided specific percentages, whereas studies siding with climatic factors provided only two percentages and, consequently, four qualitative statements. None of these 16 focus studies concluded that both stressors contributed, or contributed equally, to Lake Urmia's desiccation. Moreover, none of these focus studies is positioned at an extreme, which implies that both stressors were acknowledged despite their diverging significance among scientific findings. Second, an equal number of studies did not conduct research on the matter but were positioned in either a primarily or exclusively anthropogenic or equal place, whereas only a marginal number of papers sided with those studies implying that climate change has had the greater impact on Lake Urmia. Third, the extreme position, which implies that only one of the two stressors is relevant in this matter, was not taken by any papers on the 'natural' side, while the anthropogenic side had four studies. Lastly, after 2014, slightly fewer papers regarded the two main stressors as equal, thus moving the controversy into a more polarised debate; remember, this is what the figure suggests due to the limited categories. Based on these observations, several conclusions can be drawn and comparisons made in relation to the synthesis of Parsinejad et al. (2022).

Although Parsinejad et al. (2022) are correct in the sense that the anthropogenic impact is blamed more often as a primary cause, there is nonetheless a sufficient number of studies proving that climate change and droughts did have significant—and perhaps even a primary—impact. Non-focus studies clearly indicate—one could even say manoeuvre—anthropogenic stressors, while the neutral middle ground is less and less populated. In defence of Parsinejad et al. (2022), the exact list of studies they discussed is unknown to me, so it is possible that their synthesis, based on the 63 articles, does generate a different result. However, to move as close as possible to a similar corpus, studies on the matter were included—except for the few articles in Farsi (e.g., Pooralihossein & Delavar, 2020 see in Parsinejad et al., 2020: 5). A question that arose during the analysis concerned what Parsinejad et al. (2022: 4) specifically understood through the term 'discussed' when they synthesised the 63 articles. Did these studies contribute scientific findings, or were they literally just discussing the findings of other researchers? If the latter is the case, then the controversy, at least in the sciences, clearly runs into trouble. Even in the sciences, only objective findings, even though here they must be marked as second-degree objectivity, should count in a synthesis and not so studies that have nothing to do with the matter but choose to lean on one or the other side—or on neutral terms. This super-controversy shares quite similar characteristics with the causeway controversy, in that although both sides provide enough scientific evidence, most articles since 2014 have shifted towards, or rather against, anthropogenic stressors, while only a few articles acknowledge the controversial nature of the debate. The next similarity is the increasing number of articles that did research on this topic, especially after 2018. Beforehand, only six studies provided findings on the debate, but after 2018, 15 contributed to the debate and leant towards anthropogenic stressors as having the primary impact. The controversy remains in a state of second-degree objectivity, where only time, scientific endeavour, and the dynamics and politics of the sciences as a selection process will reveal which of the following scenarios is the most accurate: that it is a primarily human-made disaster, and thus another victim of the Aral Sea Syndrome; or that it is primarily a disaster that has been triggered by climate change, and thus a victim of collective inaction, not nationally and internationally. The trend in the sciences points at the former, but we cannot neglect scientific findings on the latter—and vice versa. This controversy will remain a heated debate for many years to come.

6.3 The Silenced Politics in Iran's Water Management

If there is one central figure that may explain why there has been such a shift in the main cause controversy towards anthropogenic impacts, the name Kaveh Madani has to be mentioned. Madani is not only a metaphorically speaking central actor in the scientific landscape, but also a literal one. Fig. 4.11 in Chapter 4.4 showed a co-authorship map of peer-reviewed articles published on Lake Urmia. Just left of a homogenous accumulation of nods we can see Madani as a globally central nod, who has been a co-author on a total of 47 articles. The cluster that is positioned adjacently to Madani is similarly characterised by a high number of co-authorships and also takes a globally central position as a whole cluster in the network. Taken together, the cluster accounts for roughly one-third of all co-authorships in the entire network (533 of 1517), which highlights the importance of this group of scientists. This cluster—led by Madani—which is apparently highly active on Lake Urmia, shares a single characteristic: all of these scientists, although Iranians in origin, all work at foreign universities, especially in the USA and Canada. For example, AghaKouchak and Alborzi were employed at the University of California, or Hassanzadeh in Montréal and finally Madani, nowadays working in Dresden, Germany—beforehand, among others, at Imperial College in London. The reason why attention turns to this cluster—but first and foremost to Madani—is to elucidate the political nature of the Lake Urmia controversy that will lead us away from putting all the blame on farmers and instead shift the focus to the government and its institutions. Moreover, based on self-conducted expert interviews—one in person, the others via email—another important group of actors will be presented that voice what is usually silenced in Iran, namely politics on water, or politics in general. It is not as if scientists or other central figures do not criticise the government for its actions or inactions. Nevertheless, there are only a few, although they are growing in number. The reason why especially Iranian scientists still residing in Iran are usually taciturn on politics lies in the fact that they fear being removed or put in jail by the government (Tamman, 2021). This statement was also expressed by several scientists during the field phase, namely that one should not mention politics while in Iran, not even, or especially, as a foreign scientist. Kaveh Madani, who was employed as deputy head of the Department of Environment in Iran, only evaded imprisonment by the Islamic Republic intelligence by luck and now lives in exile, so to speak (Tamman, 2021). One interviewee at the end of our conversation even said, 'If you post this interview on the internet, I would be arrested in Iran. I would never be able to go back home'. A veil of fear lies over most scientists in Iran, and so my hypothesis explains why Lake Urmia is treated in general apolitically, while the only way to criticise the government indirectly is by pushing the fault not onto meteorological drought or climate change but onto anthropogenic factors. Nonetheless, even in the sciences, at Lake Urmia the tide is slowly turning, and in recent years the political dimensions have been emphasised on a regular basis. The scientific narrative wishes to counter the government's narrative in order to prevent any inaction—and as such the loss of Lake Urmia. The voices positing the notion that political processes contribute to the Lake Urmia crisis deserve more attention in the following, as they provide a further perspective on the cause controversy that is usually swept under the carpet. Consequently, Madani shall be the starting point to explore this narrative as faithfully as possible.

Kaveh Madani, who was born in 1981 in Tehran, earned in 2003 an undergraduate degree in Civil-Hydraulic Engineering at the University of Tabriz, followed by a Master of Water Resources in Lund, Sweden in 2005 and finally a doctorate in Civil & Environmental Engineering at the University of California in 2009. Madani was one of the leading figures that wished to bring change to Iran's water management and who was given the opportunity to do so after he received the position of vice deputy head of Iran's Department of Environment in 2017 during the Rouhani presidency. This in itself was a controversial return (Tamman, 2021) due to his critical papers on the 'looming water crisis in Iran' (Madani, 2014) and 'Iran's Socio-economic Drought' (Madani et al., 2016), in which he accused the government's policy of being the leading cause of the water crisis in general and, among others, for the desiccation of water bodies such as Lake Urmia. The return of Madani to

Iran after 14 years of working on his scientific career in Western countries was seen as a symbolic act by Rouhani to counteract Iran's brain drain, namely to return a young, post-Revolutionary generation, trained by Western universities, back to their home country, which was facing a nationwide environmental crisis. However, the return of Madani was not welcomed by political actors—hardliners and conservatives—who accused him of espionage. The pressure put on Madani as well as his patrons curtailed the end of the aspiration to transform Iran's water management (Dehghan, 2018). After six months, Madani quit his position at the Department of Environment after the political hardliners won the internal political battle against Rouhani's environmental policy. Madani, by luck alone, evaded arrest by the Iranian Intelligence Service and has since never returned to Iran (Tamman, 2021). Although Madani has never been an expert on the Lake Urmia crisis per se, he is recognised as the leading figure in elucidating Iran's general water crisis. His views on the water crisis and its causes especially influenced those papers that touched on the political dimensions influencing Lake Urmia. Therefore, his two aforementioned papers, which had been cited 418 (Madani, 2014) and 216 (Madani et al., 2016) times, respectively, shall be summarised and analysed briefly herein. Moreover, insights shared by other experts on the matter will be included to enrich the political perspective with additional or complementary viewpoints.

In his paper 'Water management in Iran: what is causing the looming crisis?' Madani (2014) provides three major crisis drivers: population growth and its uneven spatial distribution, inefficient agriculture and mismanagement guided by a thirst for modernity or 'development', as he referred to it. The first driver, population growth, resembles exactly what Linton (2010: 198) marked as a trope that pushes the narrative of the 'Neo-Malthusian man'. The juxtaposition of limited water resources and an increasing population incapable of reacting to its changing environment 'cannot help but produce an immediate water crisis' (Linton, 2010: 198). Madani (2014: 319–320) argues that economic development during the Pahlavi era, coupled with the population policy exercised by the Islamic government after 1979, caused significant population growth. From 1979 to 2014, Iran's population more than doubled from 35 to 77 million, which by default put pressure on renewable fresh water resources. However, the first driver turns out to be more than just a simple Neo-Malthusian argument. Although population growth is regarded as a major driver, its negative impact on fresh water resources is multiplied by the socio-political context. Iranians are depicted on average as excessive water consumers, who tend to move to highly water-demanding urban areas despite their limited water resources. As water demand increases, cities are in dire need of water transfer projects to sustain themselves. The situation is exacerbated first by the spatial concentration of these urban areas, second by their power over rural areas in terms of water competition and third by the population politics of the government, who wish to boost population growth. Clearly, Madani is concerned with more than mere population growth and its distribution. Much more visible are the political processes that really make population growth a problem, namely the power imbalance between urban and rural areas, the political drive to continue population growth despite the unfortunate water crisis and infrastructure projects that exacerbate rather than ameliorate unsustainable practices with limited water resources. In relation to this first point, this first driver, everything is just drenched in politics, and it is already encroaching on the two other points: agriculture and poor water management. Before I go into these stressors in more detail, however, it is worth adding to the above views expressed in my interviews with experts in the media and in studies. For this purpose, the population growth argument and its distribution as well as the political dimensions that play a role here should be examined in more detail. According to Interviewee A, the government is focusing its water resources on Iran's Central Plateau attempting to supply cities and sensitive industries in this area, which is mainly populated by the major ethnicity of Iran, the Farsi. Sensitive industries are foundries such as iron or copper factories that, much like the increasing urban areas, have to be provided with large amounts of water, allocated and distributed by dams, pipelines and desalination plants, all of which are veiled as government projects to supply drinking water. Interviewee A provided the following concrete examples:

> 'In Yazd, there is an iron foundry, it produces about 1,000 tonnes of iron per day, and keep in mind they need 25,000 m3 [of water] to produce only one tonne of iron. They are getting this water through a pipeline from Zyahineh-Rud in Isfahan. The length of this pipeline is 333 kilometres. They [the government] are bringing in this, but they are not saying outspokenly that, "We are bringing this water for the purpose of producing iron or whatever. We need this water for drinking purposes." Of course, a

big proportion of this water is being used in the domestic sector, but this is not the only purpose. This you can call a sort of covert purpose'.

The high water demand of urban areas, especially for industries, combined with a proliferating technical infrastructure, however, was regarded by the interviewee as being secondary in relation to population growth. Against this backdrop, the Malthusian theory was brought up on the matter and used as an explanation:

> 'The worst is something else. The worst effect is the cancerous population growth in the central plateau ... You can build up on the economic theory of Thomas Malthus. So, if you give water to [the people], to a city like that [e.g., Yazd in the central Plateau], from an unsustainable resource, like water transfer or dam construction or desalination, the population will increase automatically. An increase in combination will be exponential. So you can't keep pace with population growth all the time. Sooner or later, it will reach a point where you have to succumb, you have to say, "Okay, I give up. I'm completely out of water, I cannot provide you with more water." Exactly at that point, the hydraulic collapse will happen'.

Given this national scenario presented by the experts, these aforementioned phenomena are also regularly correlated with Lake Urmia. First of all, population growth is also mentioned by other researchers as a driver of Lake Urmia's desiccation, as well as urbanisation. The 'phenomenal' (Hashemi, 2012: 136) or 'sharp' (Zarrineh & Abad, 2014: 41) population growth in the Lake Urmia basin has had a 'detrimental role in the decline' of the lake (Alipour & Olya, 2014: 4). From 1976 to 2010, the population in the basin increased by 121.5% (2.6 million to 5.9 million) (Hashemi, 2012: 136). However, the statistical assessment for 2010 was not based on statistical data but on estimates made by stakeholders. Disturbingly, the exact population nowadays remains a twisted affair, as it varies between 7.9 (Feizizadeh et al. 2023) to 6.5 million (Shadkam, 2017: 55; Haghighi et al., 2018: 3596), to finally 5.12 million in 2016 according to the ULRP (2018: 8). The ratio between the rural and urban population remains unknown, and yet the ULRP (2018: 40) provides a statistic on urban water consumption in East and West Azerbaijan. In this context, one must be aware that Madani (2014: 319) argued that Iranians are depicted as excessive water consumers, consuming twice the world's average (250 l/c/d) and in some urban areas even significantly more than that (400 l/c/d). If the statistics from 2011 to 2016 provided by the ULRP (2018) are trustworthy and 250 l/c/d is perceived as excessive water consumption in Iran, even then only two cities (Khosroshahr & Basmenj) would fall under this category, whereas per capita consumption in the largest cities Tabriz (1.77 million) and Urmia (1.04 million) is estimated at around 175 litres per day. The criticism of urbanisation, population growth and excessive water consumption is missing a vital point, or rather a normative perspective: What is the minimum threshold of water consumption for people, and how is this amount of water compatible with lake restoration efforts? To provide another viewpoint—one of water ethics—Feitelson (2012: 54) discussed the minimum amount of water necessary to live a dignified live capable of human development. The threshold varies significantly from 50 l/c/d (Gleick, 1996), to 274 l/c/d (Falkenmark, 1986), while Chenoweth (2008) calculated 135 l/c/d and Feitelson (2012: 54) considers 164 l/c/d in semi-arid and arid regions—due to the higher evaporation rates—'as a normative domestic water need (including municipal use)'. To make things clear, even among researchers there is no consensus what the minimal threshold should be for human water needs, let alone the fact that Falkenmark's (1986) or Feitelson's suggestions would mark urban water consumption in the Lake Urmia basin rather acceptable or even slightly to low in some cities. In this case, Madani's (2014) criticism seems to be less applicable than for other urban areas such as Tehran.

However, at Lake Urmia, population growth and urbanisation are primarily correlated in the literature with increasing water and food demands, and hence expansion in the agricultural sector that places even more pressure on limited fresh water resources.

A further distinction that has to be drawn to the national scenario is the fact that Lake Urmia lies in a peripheral region of Iran, where industries are mentioned in the literature as crisis drivers (e.g., Lotfi, 2012: 89), and yet this is rarely the case and therefore negligible as a driving force at Lake Urmia.

A factor that does play a role in the Lake Urmia basin is inter-basin water transfer. Tabriz, for example, receives much of its water from West Azerbaijan, including the Zarrined Rud, from a pipeline constructed in 1999 providing 157 million m^3 to the city (Alizadeh-Choobari et al., 2016: n.p.). However, Tabriz supposedly suffered from water shortage and hence established a second pipeline, which according to Agh during our meeting at Lake Urmia in 2019 was installed and caused civil unrest in those regions from where the water was diverted. Moreover, Agh commented that the amount of water transferred by the second pipeline was negligible and labelled the protests irrational, while the first pipeline's water transfer to Tabriz was regarded as 'huge' and contributing the socio-economic drought in the basin by Alizadeh-Choobari et al. (2016: n.p.). The inequity of water distribution is a topic less discussed in the science at Lake Urmia, yet Ženko and Menga (2019: 10) argued that upstream and urban water demand facilitated by hydraulic infrastructure generated water scarcity in downstream areas, which seem to be at the losing end of water distribution. Khalyani et al. (2014), all from US-American universities—openly criticise the construction of the first pipeline, established from Zarrineh-Rud to Tabriz, which uses apparently 300 Mm^3 of water for domestic, industrial and agricultural purposes in the area. The Kurds, however, have been deprived of their water, and even more so:

> 'The water supply and filtration functions that the ecosystem provides are critically important to the Kurdish populations, and represent a spatial distribution of water resources that also maintains Lake Urmia and its southern wetlands, supporting biodiversity, agriculture, and tourism. Had the importance of these water resources to *Kurdish communities* been understood as *equally important* to the *health of Lake Urmia*, the long-term negative consequences of the water infrastructure projects *might have prevented* the construction'. (Khalyani et al., 2014: 5; emphasis added)

The moral and political implications here are clear: if decision-makers would have included the needs of humans, in this case the Kurdish, and of Lake Urmia, the pipeline would not have been constructed. Nonetheless, both the Kurds and Lake Urmia seem to be losing out in terms of the distribution of river water against urban areas or Tabriz in particular.

In this case, Madani's (2014) general rule of thumb does apply, in that that urban areas superimpose their water needs over rural areas. Even so, the impressions from my field visit may allow me to share another perspective. Indeed, downstream areas did experience a lack of water for irrigation purposes in both West Azerbaijan and East Azerbaijan—in a few cases even potable water, due to bad water quality. Downstream farmers complained about dammed rivers, whose water they could only use when dams were opened. Some farmers added that they were prohibited from using river water even during dam openings, as it was meant exclusively for Lake Urmia; in both cases, they were forced to use groundwater and hence drill illegal wells. Furthermore, other farmers included the increasing conflicts among up- and downstream farmers and even villages, whereby the former used more water to irrigate their water-intensive crops. The water distribution system is designed in a way that prioritises the needs of users in upstream areas. In this regard, I am in agreement with the findings of Ženko and Menga (2019). A slightly different impression of mine is that cities did provide rural areas with, for example, potable water. Yet, in one case, the villagers completely distrusted the water transferred from the city through a pipeline and refused to use it, which could be classed as a 'silent act of resistance' (Benjaminsen & Svarstad, 2019: 393). Differences in terms of access to water, however, are evident and drive a wedge between rural and urban areas, with the latter gaining more and more control over water resources. Interviewee A made an interesting point on this subject:

> 'Nowadays, there is no [mutual] relationship [...] between the villages and the cities. Villages are completely dependent on cities, and this is sort of a one-way relationship. Even drinking water comes from the city because of drought and the mismanagement of water resources in Iran'.

The next driving force in the Iranian water crisis on Madani's (2014) list was inefficient agriculture. Again, this is a topic well renowned and acknowledged in the sciences at Lake Urmia. Nonetheless, it mostly does not explain why the agricultural sector is in the state that it is, which is mostly perceived as archaic or outdated, and hence economically inefficient and water-inefficient. Madani (2014) and Interviewee A shared a common reason for this phenomenon: political disinterest. More precisely, after Iran transformed during the twentieth century to an oil-based economy, the modernisation of agricultural practices was neglected by the government (Madani, 2014: 320) despite its ambition to be food self-sufficient (Dalby & Moussavi, 2016: 7–8). The government pushes the agricultural sector to increase food production while failing to subsidise and modernise its methods. Moreover, the economic situation, including the trade embargo, puts even more pressure on farmers, thereby increasing their socio-economic hardship. Pouladi et al. (2021: 5) take off some of the guilt that has been put on farmers, as they '[i]nevitably [...] try to compensate for their unfavourable living and economic conditions by exploiting land and water resources'. Even Issa Kalantari, the head of ULRP, in an interview put the blame not on farmers but on the government's inability to react to Lake Urmia's situation properly (*Tehran* Times, 2016). Agriculture lays the socio-economic foundation of rural areas and provides sufficient job opportunities (Pouladi et al., 2021: 5). Yet, this sector is missing out on financial investment because the government has no interest in spending its 'oil money' to improve irrigation systems but rather on:

> 'bureaucratic integration [and] on improving the military, military communities, some infrastructure across the country like roads, airports. They prefer to produce the foodstuff within the country, not import food stuff from other countries as virtual water. That is why they are encouraging agriculture'. (Interviewee A)

Nonetheless, at the same time, self-sufficiency represents a major pillar of Iran's political ideology, as it provides political stability. In other words, while agriculture may not receive sufficient investments to install water-efficient irrigation technologies, it is synonymous with political stability, which allows the government to maintain its powerful position. The government hence has no interest in enforcing measures on farmers that would increase socio-economic pressure as well as the likelihood of civil unrest. Therefore, the government tolerates farmers' detrimental behaviour for the sake of political stability (Madani, 2014: 320; Salimi et al., 2019: 631). Moreover, decision-makers facilitate water exploitation through populist measures (e.g., the proliferation of wells) that generally distort water as 'an open access, free of charge resource' (Schmidt, 2021: 189). The government's nonchalant stance on water resources is contrasted, for example, by an interview with Expert-2, who stated that 'water is an economic good and has its own value'. Nevertheless, its value is misrepresented, as the price for irrigation water equals 'only 3% of the Value of Total Product (VTP) per hectare in modern water canals' (Expert-2). This confirms the findings of Pouladi et al. (2021: 4) that water is literally undervalued by the government. As a result, the blame game by some scientists has turned from pinpointing farmers—perceived as victims of their socio-political context—to the central government, which is made responsible for Iran's general water bankruptcy (Madani et al., 2016) and hence Lake Urmia's desiccation. This, in part, is due to Iran's food self-sufficiency agenda, which is marked by 'decades of irresponsible and ideologized policies' (Mesgaran & Azadi, 2018: 5) causing severe environmental degradation and water exploitation. After population growth, urbanisation, water transfers, inefficient agriculture and government policies, I finally arrive at Madani's (2014: 321–323) last point, namely 'mismanagement and thirst for development'.

6.4 The Tyranny of Small Decisions and Myopia

In short, this driver explains how water is simultaneously an open access resource and highly desirable. The prioritisation of short-term political and economic gains, often achieved through the exploitation of natural resources, can lead to unsustainable practices due to a neglect of long-term environmental and societal consequences. For Madani (2014: 321), Lake Urmia desiccated due to anthropogenic factors, but more precisely 'selfish and uncoordinated upstream development activities by three provinces in a competitive environment'. Over the years, some studies have tacitly validated this—once again—rather socio-political argument. Salimi et al. (2019: 630) reported on the difficulties inherent in the centralised, top-down water management system in Iran, whose actions are often translated by local agencies to the detriment of sustainable solutions. Due to diverging interests of stakeholders, more water is used than ever before. Expert-1, for example, addressed the Ministry of Energy and Water and the Ministry of Agriculture in this context, arguing:

> 'Frankly, coping with water scarcity contradicts the main tasks of these organisations! The benefits of these agencies have actually been established in more water consumption!'

Similarly, Expert-2 explained in more detail:

> 'In relation to water, there is a conflict of interest between the two organisations. The Agricultural Organisation seeks to develop the agricultural sector, and the Water Organisation seeks to make a profit by selling water to farmers. In order to create co-operation and coordination between these two organisations, they need to give up part of their profits. In the current situation, such a thing is not observed'.

As such, the Lake Urmia crisis is also the result of uncoordinated and conflicting stakeholder interests. This situation is exacerbated by the fact that Ahmadinejad changed water management 'from watershed boundaries to political (provincial) boundaries' (Madani, 2014: 321), leaving the watershed as a whole prone to unsustainable profit-maximisation, conflicts of interest and corruption. This playing field allows, in Pouladi et al., and and's (2021: 9) view, the proliferation of individual and egocentric decisions—known as the 'tyranny of small decisions'—and as such mark the root cause of Lake Urmia's crisis. Farmers obviously are key actors in this tyranny, but top-down decision-making processes lack co-operation and mutual learning between these local farmers and state officials and water agencies (Pouladi et al. 2019: 10). Furthermore, farmers do make the government accountable for the Lake Urmia crisis on multiple levels. First, the government is by constitutional law the most powerful decision-maker on water resources, and its responsibility is therefore to manage the situation to the best of its abilities (Abbaszadeh et al., 2020: 1). In fact, Article 45 in Chapter IV of the Iranian Constitution confirms this point, at least on a legal note: 'Public wealth and property, such as [...] seas, lakes, rivers and other public waterways [...] shall be at the disposal of the Islamic government for it to utilize in accordance with the public interest' (Iranian Constitution n.d.). Interviewee A, however, commented that water, along with all other natural resources, does not belong to the nation. According to the eighth Sura in the Coran, called Al-Anfāl or simply Anfāl, natural resources are interpreted by Shiite clergymen as war booty, which in fact belong to the Shiite Imam or, in his absence, the Supreme Leader of Iran. Hence, 'if all water resources belong to the Islamic government, it can make whatever decision about water resources it wishes to make', suggested Interviewee A, Article 45.

Farmers at Lake Urmia, on the other hand, perceive themselves as victims at the very tail end of water resource management with no power whatsoever. From their point of view, the government has disturbed traditional water management and failed to exercise and make use of its power, while corrupt local governors and agents compete over water resources. Farmers have a sense of powerlessness and exclusion from the central government policy, which neglects the interests of other ethnicities such as the Azeri and especially the Kurds (Ženko & Menga, 2019; Ženko & Ulezic, 2019). These factors drive farmers into a passive, lethargic state of mind, which paradoxically gives them legitimation to continue their—in some cases—water overuse despite the constant decline of water resources (Abbaszadeh et al., 2020: 1–2). Farahmand et al. (2023: 13) critically comment on some farmers' dreadful behaviour towards water resources and those who blame high inflation, the trade embargo, other farmers or the government for environmental issues, but not themselves. Indeed, one could interpret this as the tyranny of victimhood and ignorance, as numerous scientists register and complain about the low level of awareness on environmental issues, which allows, for example, farmers to believe that there is enough water in the basin. This attitude is polemically addressed by Farahmand et al. (2023: 13), who write that 'farmers do nothing even if the lake disappears from the earth'. Yet, their aggressive use of water resources (Madani, 2014: 318 Madani et al., 2016: 1012), which should be met by equally 'aggressive restoration policies' according to Alborzi et al. (2018: 10), is not the only aggression enacted on water in the Lake Urmia basin. Similarly, as discussed above, 'aggressive agricultural policies' (Ženko & Menga, 2019: 7) and 'aggressive hydro-economic development' (AghaKouchak et al., 2015: 309) have allowed farmers to pursue their unsustainable activities.

The Islamic Revolution and its thirst for development has enacted Iran's hydraulic mission (Madani, 2014: 317). The 'Fair' or 'Just Distribution Water Law', introduced into the Constitution in 1983 (JICA, 2016: 35), and the aforementioned Article 45 propelled the hydraulic mission, which in a paradoxical contrast to the Pahlavi era resulted in the construction of more dams in Iran than ever before (Nabavi, 2017: 66). The 'dam construction syndrome', as labelled by Interviewee A, produced staggering—but probably exaggerated—figures: although Iran's geographical area covers only 3% of the Asian continent, 40% of it (roughly 650 dams) has been built on (Interviewee A). Dams symbolise the modernisation paradigm that is a characteristic of the hydraulic mission for Middle Eastern countries in general (Schmidt, 2021: 189). Despite the development of the agricultural sector, these 'state-sponsored damming projects' have been accused by sciences, the media and the public as being a main cause of Lake Urmia's desiccation (Ranjbar, 2022: 2). In 2015, at a conference of the Unrepresented Nations & Peoples Organization (UNPO), a spokesperson for the South Azerbaijan Student Movement, from Lake Urmia, argued that more than 100 dams had been constructed in the basin and the government did not intend to change its dam policy in favour of Lake Urmia; even worse, the government was purposely draining it. The goal of the government, so the spokesperson—and one of my interviewees' in the field—noted, was to use salt from Lake Urmia, 'incredibly rich in Uranium', for its atomic programme (AZOH, n.d.). However, at this point, I wish to label this a conspiracy theory, because other than these sources, I could not find any papers on the matter. This might explain why the UNPO did not publish this statement on its homepage. Nonetheless, even if this conspiracy is wrong, the public perceives this excessive dam construction by the government as a violent act against Lake Urmia and as such as a political act against the socio-economic well-being, as well as the human rights of the basin's population. The government pursued its dam policy with the argument that climate change rather than anthropogenic factors was causing Lake Urmia's demise (Ranjbar, 2022: 2), although Chitichian, head of the MoE, admitted in 2016 on public television that dams were indeed a significant issue (AZOH, n.d.). Regardless, dams have been constructed under the gaze of the quasi-military Revolutionary Guard in the past 30 years, an association (Ranjbar, 2022: 6) directly related to the Supreme Leader. The real profiteers in the context of this dam construction syndrome, probably including the Lake Urmia basin, is the so-called water mafia according to Interviewee A. This group of actors may be the stakeholders marked by Mesgaran & Azadi (2018: 23) as corrupt and responsible for the construction of a 'large number of unnecessary dams from the public budget while overlooking the development of several hundred thousand illegal water wells'. According to Interviewee A, the water mafia 'does nothing to stop dam construction syndrome' and tries to maintain the modernist paradigm for its own profit. Furthermore:

> '[The water mafia is] scattered across the political fabric. You can find members in many parts of the Iranian polity, but all of them are connected somehow. They are looking out for each other's profit and interests. For example, some of them can be found in the Ministry of Energy, and some of them are in the Revolutionary Guard in the Shrine of Imam, the Ministry of Oil. They work in tandem, they co-operate with each other. Because their interests and profits are knitted together, are tied together. That is why they are supportive of each other; they are actually advocating each other's policies and decisions'. (Interviewee A).

The dam policy in Iran intensifies the general mistrust between the government and local dwellers at Lake Urmia. Protests that erupted in 2011 in Tabriz and Urmia highlighted political tensions in this region, which were answered brutally by the police (Dehghan, 2011). Ranjbar (2022: 12), however, criticises 'English-language media' who reduced the protests to either an environmental or a political issue while remaining ignorant on the 'deep rationality' between Lake Urmia and the local population. The environmental movement was a political strategy by protesters to minimise the risk of openly challenging the government and using 'green' coded language to express their socio-political dissatisfaction, with Lake Urmia being a central pillar of their well-being (Ranjbar, 2022: 12). In the aftermath of the protests in 2011, a local MP tried to depoliticise the issue by saying:

> 'In my view, the issue [of Lake Urmia] should not be seen as a security issue and it should not be politicised. It is a social and environmental issue which we can rescue, and it can be solved by the human'. (Dehghan, 2011).

However, in 2022, newly rising protests in Urmia indicated that the situation was as socio-politically tense as before; protesters shouted slogans such as 'Lake Urmia is dying, parliament orders its killing' (VOA, 2022).

It seems as if both the local population and scientists in Iran have limited space for criticism, which allows them to more or less circumvent any politically sensitive issues by either pointing at the dramatic state of Lake Urmia or, what sciences do, blaming farmers and the way they exploit water, which the government similarly tiptoes around. In this regard, the influential papers of Madani next to the aforementioned studies display the very political nature of the Lake Urmia crisis. Yet, Madani should be viewed as probably the leading figure in the active campaign against state-led ignorance towards water resources. In the context of politics, it is evident that certain actors are able to gain from the hydraulic mission, exemplified by the 'water mafia', while others remain constrained. This observation is supported by Madani et al. (2016: 1006), who highlighted the limited responsibility assigned to the Department of Environment in the protection of Iran's environment. Depoliticising and dissocialising the Lake Urmia crisis threatens to completely distort hydro-social relations and limit, as we shall see, any solution highly impractical and equally controversial.

6.5 Conclusions: Lake Urmia is a Casualty of Human Error

The controversy over the prima causa of Lake Urmia's desiccation, or mutation, is not settled yet. Nevertheless, analysing the different positions on the matter has revealed several insights. First, similar to the causeway controversy, studies that did try to identify the main drivers of the crisis are relatively scarce, and their findings often diverge. However, they usually do not drift apart as drastically; that is to say, most studies agree that the demise of Lake Urmia is the result of interlocking mechanisms—both exogenous and endogenous. Interestingly—and again, reminiscent of the causeway controversy—since 2014, most scientists have sided with what AghaKouchak labelled the 'anthropogenic drought' (Stone, 2021: 445). For the majority of scientists, anthropogenic factors in situ are the main reason, including dam construction, the expansion of agriculture, the proliferation of illegal wells, myopic and conflicting water management, unsustainable urban water use and the farmers' choice of water-intensive crops. Yet, at the same time, other studies that tend to attribute exogenous or 'natural' causes as the main contributors to the lake's mutation, cannot be ignored, even though, for example, the synthesis report by Parsinejad et al. (2022) tried to short-circuit the debate artificially—both by ignoring certain studies (such as Schulz et al., 2020) and mixing or combining numbers, or rather percentages, in a confusing fashion. This is not to say that the scientific push to hold human actors—especially farmers—accountable for the disaster would be illegitimate, but it is not founded on facts in the sense of traditional objectivity (first-degree objectivity) but second-degree objectivity.

Apart from these insights, the discourse has also shown a perhaps much too neglected aspect of the Lake Urmia crisis: its political nature and an extended understanding of responsibility. As Madani explained in an interview, scientists in developing countries such as Iran 'have to warn about climate change and make sure climate change is recognized as a serious threat, but at the same time make sure that governments don't misuse climate change' (Tamman, 2021). In this sense, the responsibility is attributed not only to the direct factors influencing the flow of water into Lake Urmia, but also indirect, remote influences. This means that the water crisis neither begins nor ends at the diesel-fuelled water pumps, the dams and their water reservoirs, water taps, pipelines or the farmers' decision-making processes but at conferences, paperwork on desks, political ambitions and intentions and historical events that still have a significant impact on the lake (take, for example, the White Revolution, the dissolution of the *buneh* system, the technocratic hydraulic mission). The responsibility is thus shared by an extended circle of actors that are indeed political, although most papers seem—for several reasons—to neglect these aspects. Therefore, the primary culprits are usually identified mostly in situ, whereas remote factors are understudied.

This also shapes our understanding of water. Assuredly, water is matter, but it is also a political matter, and the decisions and directions that are being taken change the flow of water, and in this case, they redirect rivers and reduce water tables drastically. Either action or inaction has consequences, and the mutation of Lake Urmia shows how different actions are interrelated and lead to unintended consequences (ignoring the conspiracy theories, namely that Lake Urmia was intentionally desiccated). I wish to continue to analyse and discuss this aspect in the following chapter.

References

Abbaszadeh, M., Banifatemeh, H., & Nikdel, N. (2020). A study of Farmer's Perception of Urmia Lake (UL) crisis with grounded theory approach. *Journal of Applied Sociology, 31*(2), 1–4.

AghaKouchak, A., Norouzi, H. & Madani, K. et al. (2015). Aral Sea syndrome desiccates Lake Urmia: Call for action. *Journal of Great Lakes Research 41*(1), 307–311.

Alborzi, A., Mirchi, A. & Moftakhari, H. et al. (2018). Climate-informed environmental inflows to revive a drying lake facing meteorological and anthropogenic droughts. *Environmental Research Letters 13*(8), 084010.

Alipour, H., & Olya, H. G. T. (2014). Sustainable planning model toward reviving Lake Urmia. *International Journal of Water Resources Development, 31*(4), 519–539.

Alizadeh-Choobari, O., Ahmadi-Givi, F., Mirzaei, N., & Owlad, E. (2016). Climate change and anthropogenic impacts on the rapid shrinkage of Lake Urmia. *International Journal of Climatology, 36*(13), 4276–4286.

Azarnivand, A., Hashemi-Madani, F. S., & Banihabib, M. E. (2015). Extended fuzzy analytic hierarchy process approach in water and environmental management (case study: Lake Urmia Basin, Iran). *Environmental Earth Sciences, 73*(1), 13–26.

AZOH. (n.d.). *Speech in European Parliament seminar for Urmia Lake: AZOH spokesman.* https://azoh.info/index.php?option=com_content&view=article&id=28994:speech-in-european-parliament-seminar-for-urmia-lake-azoh-spokesman&catid=33:english. Accessed 1/27/2024.

Benjaminsen, T. A. & Svarstad, H. (2019) Political ecology. In: *Encyclopedia of ecology* (pp. 391–396). Elsevier.

Chaharborj, A. J. (2014). Evaluation of Iran's Performance in relation to the environmental problem of the Urmia Lake from the Perspective of International Environmental Law. *Trends in Life Sciences, 3*(3), 1.

Chenoweth, J. (2008). Minimum water requirement for social and economic development. *Desalination, 229*(1–3), 245–256.

Collins, H., & Pinch, T. (1993). *The Golem: What everyone should know about science.* Cambridge University Press.

Dalby, S., & Moussavi, Z. (2016). Environmental security, geopolitics and the case of Lake Urmia's disappearance. *Global Change, Peace & Security, 29*(1), 39–55.

Danesh-Yazdi, M., & Ataie-Ashtiani, B. (2019). Lake Urmia crisis and restoration plan: Planning without appropriate data and model is gambling. *Journal of Hydrology, 576*, 639–651.

Dehghan, S. K. (2011). Iranian greens fear disaster as Lake Orumieh shrinks. *The Guardian,* 2011. https://www.theguardian.com/world/2011/sep/05/iran-greens-lake-orumieh-shrinks. Accessed Augest 9, 2024.

Dehghan, S. K. (2018) Top scientist leaves Iran after crackdown on environmentalists. *The Guardian,* 2018. https://www.theguardian.com/environment/2018/apr/18/kaveh-madani-scientist-leaves-iran-after-crackdown-environmentalists. Accessed Augest 9, 2024.

Falkenmark, M. (1986). Fresh Water: Time for a Modified Approach. *Ambio, 15*(4), 192–200.

Farahmand, H., Tajrishy, M., Isaai, M. T., Ghoreishi, M., & Mohammadi, M. (2023). *How do farmers' perceptions and attitudes toward agricultural water consumption behaviors can lead to unsustainability; evidence from Mahabad plain.* Lake Urmia.

Feitelson, E. (2012). What is water? A normative perspective. *Water Policy, 14*(3), 386–417.

Feizizadeh, B., Blaschke, T., Nazmfar, H., Akbari, E., & Kohbanani, H. R. (2013). Monitoring land surface temperature relationship to land use/land cover from satellite imagery in Maraqeh County, Iran. *Journal of Environmental Planning and Management, 56*(9), 1290–1315.

Gleick, P. H. (1996). Basic Water Requirements for Human Activities: Meeting Basic Needs. *Water International, 21*(2), 83–92.

Haghighi, A. T., Fazel, N., Hekmatzadeh, A. A., & Klöve, B. (2018). Analysis of Effective Environmental Flow Release Strategies for Lake Urmia Restoration. *Water Resources Management, 32*(11), 3595–3609.

Hashemi, M. (2012). *A social-technical assessment framework for integrated water resources management (IRWM) in Lake Urmia Basin, Iran.* Doctor of Philosophy Thesis. Newcastle University.

Hassanzadeh, E., Zarghami, M., & Hassanzadeh, Y. (2012). Determining the Main Factors in Declining the Urmia Lake Level by Using System Dynamics Modeling. *Water Resources Management, 26*(1), 129–145.

Hosseini-Moghari, S.-M., Araghinejad, S., Tourian, M. J., Ebrahimi, K., & Döll, P. (2020). Quantifying the impacts of human water use and climate variations on recent drying of Lake Urmia basin: The value of different sets of space-borne and in situ data for calibrating a global hydrological model. *Hydrology and Earth System Sciences, 24*(4), 1939–1956.

Iranian Constitution. (n.d.). https://ecnl.org/sites/default/files/files/2021/IranConstitution.pdf. Accessed Augest 13, 2025.

JICA. (2016). *The Data Collection Survey on Water Supply Sector in the Islamic Republic of Iran - Final Report.* https://openjicareport.jica.go.jp/pdf/12266953.pdf. Accessed Augest 16, 2025.

Kanani, R., Fard, A. F., Ghorbani, M. A., & Dinpashoh, Y. (2020). Analysis of the role of climatic and human factors in runoff variations (case study: Lighvan River in Urmia Lake Basin, Iran). *Journal of Water and Climate Change, 11*(1), 291–302.

Khalyani, A. H., Mayer, A. L., & Norman, E. S. (2014). Water Flows Toward Power: Socioecological Degradation of Lake Urmia, Iran. *Society & Natural Resources, 27*(7), 759–767.

Khazaei, B., Khatami, S., Alemohammad, S. H., et al. (2019). Climatic or regionally induced by humans? Tracing hydro-climatic and land-use changes to better understand the Lake Urmia tragedy. *Journal of Hydrology, 569*, 203–217.

Latour, B. (2004). *Politics of Nature: How to Bring the Sciences into Democracy.* Harvard University Press.

Latour, B. (2013) *An Inquiry into Modes of Existence: An anthropology of the moderns,* 1st Harvard University Press paperback ed., 1st printing. Harvard University Press.

Linton, J. (2010). *What is water?: The history of a modern abstraction.* UBC Press.

Lotfi, A. (2012). Lake Uromiyeh A Concise Baseline Report.

Madani, K. (2014). Water management in Iran: What is causing the looming crisis? *Journal of Environmental Studies and Sciences, 4*(4), 315–328.

Madani, K., AghaKouchak, A., & Mirchi, A. (2016). Iran's Socio-economic Drought: Challenges of a Water-Bankrupt Nation. *Iranian Studies, 49*(6), 997–1016.

Mesgaran, M. B. & Azadi, P. (2018). *A National Adaptation Plan for Water Scarcity in Iran.*

Nhu, V.-H., Mohammadi, A., Shahabi, H., et al. (2020). Monitoring and assessment of water level fluctuations of the Lake Urmia and its environmental consequences using multitemporal landsat 7 ETM+ images. *International Journal of Environmental Research and Public Health, 17*(12), 4210.

Nimmo, R. (2011). Actor-network theory and methodology: Social research in a more-than-human world. *Methodological Innovations Online, 6*(3), 108–119.

Nouri, H., Mason, R. J., & Moradi, N. (2017). Land suitability evaluation for changing spatial organization in Urmia County towards conservation of Urmia Lake. *Applied Geography, 81*, 1–12.

Parsinejad, M., Rosenberg, D. E., Ghale, Y. A. G., et al. (2022). 40-years of Lake Urmia restoration research: Review, synthesis and next steps. *The Science of the Total Environment, 832*, 1.

Pouladi, P., Badiezadeh, S., Pouladi, M., et al. (2021). Interconnected governance and social barriers impeding the restoration process of Lake Urmia. *Journal of Hydrology, 598*, 1.

Ranjbar, A. M. (2022). The greening of human rights in Iran: Lake Orumiyeh, human rights, and environmental justice. *Environment and Planning e: Nature and Space, 3*(5), 1.

Salimi, J., Maknoon, R., & Meijerink, S. (2019). Designing Institutions for Watershed Management: A Case Study of the Urmia Lake Restoration National Committee. *Water Alternatives, 12*(2), 609–635.

Schmidt, M. (2021). *The Vanishing of Iran's Lake Urmia* (pp. 171–190). Brill, Leiden.

Schulz, S., Darehshouri, S., Hassanzadeh, E., Tajrishy, M., & Schüth, C. (2020). Climate change or irrigated agriculture—What drives the water level decline of Lake Urmia. *Scientific Reports, 10*(1), 236.

Shadkam, S. (2017). *Preserving Urmia Lake in a Changing World: Reconciling anthropogenic and climate drivers by hydrological modelling and policy assessment*. Wageningen University.

Stone, R. (2021). After revival, Iran's great salt lake faces peril. *Science (New York, n.y.), 372*(6541), 444–445.

Tamman. (2021). *The Hot List: Meet Kaveh Madani, the climate expert who fled Iran, April 27.* https://www.reuters.com/investigates/special-report/climate-change-scientists-madani/?utm_medium=Social&utm_source=twitter. Accessed May 12, 2022.

Tehran Times. (2016). *'Restoring Lake Urmia costs $7b, neglecting it incurs $1tn loss'*. https://www.tehrantimes.com/news/407713/Restoring-Lake-Urmia-costs-7b-neglecting-it-incurs-1tn-loss. Accessed November 8, 2024.

Tourian, M. J., Elmi, O., Chen, Q., Devaraju, B., Roohi, S., & Sneeuw, N. (2015). A spaceborne multisensor approach to monitor the desiccation of Lake Urmia in Iran. *Remote Sensing of Environment, 156*, 349–360.

ULRP. (2018). *Urmia Lake: Lessons and challenges*.

United Nations Environmental Programme (UNEP) & Global Environmental Alert Service (GEAS). (2012). The Drying of Iran's Lake Urmia and its Environmental Consequences. *Environmental Development, 2*, 128–137.

UNPO. (2015). *Conference at European Parliament Addresses Lake Urmia's Alarming Deterioration*. https://unpo.org/conference-at-european-parliament-addresses-lake-urmia%ef%bf%bds-alarming-deterioration/. Accessed January 27, 2024.

Venturini, T. (2010). Diving in magma: How to explore controversies with actor-network theory. *Public Understanding of Science, 19*(3), 258–273.

Venturini, T. (2012). Building on faults: How to represent controversies with digital methods. *Public Understanding of Science (Bristol, England), 21*(7), 796–812.

VOA. (2022). Iran Arrests Several After Protests at Drying Lake. *Voice of America (VOA News)*, 2022. https://www.voanews.com/a/iran-arrests-several-after-protests-at-drying-lake/6662459.html. Accessed Augest 9, 2024.

Zarghami, M., & AmirRahmani, M. (2017). A system dynamics approach to simulate the restoration plans for Urmia Lake, Iran. In A. Matsumoto (Ed.), *Optimization and dynamics with their applications* (pp. 309–326). Springer.

Zarrineh, N., & Abad, M. A. N. (2014). Integrated water resources management in Iran: Environmental, socio-economic and political review of drought in Lake Urmia. *International Journal of Water Resources and Environmental Engineering, 6*(1), 40–48.

Ženko, M., & Menga, F. (2019). Linking Water Scarcity to Mental Health: Hydro-Social Interruptions in the Lake Urmia Basin, Iran. *Water, 11*(5), 1.

Ženko, M. & Ulezic, S. (2019). *The unequal vulnerability of Kurdish and Azeri minorities in the case of the degradation of Lake Urmia, Iran.*

CHAPTER 7

Restoration Efforts: The (Moral) Weight of Lake Urmia

Fear is a major driving force of Lake Urmia's hydro-social metamorphosis. However, it is necessary to differentiate this point here, because in this case it is not the fear of articulating political opinion on the matter but one that projects fear of the coming future. While much of this thesis focused on elucidating differences between human actors, identifying consensus remains, nonetheless, an important part of controversy mapping. The common denominator that scientists, the public and politicians share is fear, namely that Lake Urmia is going to suffer the same fate as the prominent Aral Sea. Hence, the term 'Aral Sea Syndrome' was designated to describe the situation at Lake Urmia and as such determines the most probable outcome of the crisis: Lake Urmia will disappear, or at least shrink, causing a cascade of environmental degradation and socio-economic hardship for the people of the basin and probably even beyond the borders of the watershed. The following chapter attempts to summarise the most feared negative impacts of the crisis, some of which have already occurred.

Although stakeholders agree on why it is necessary to save Lake Urmia, they do not find common ground on how to do it, because, as this chapter will show, the solutions are always the result of different moral implications, ethical positionalities and therefore different values. The visions of what ought to be vary drastically depending on what the actors believe to be the best likely outcome for the given situation. However, what

R. Gonda, *Lake Urmia's Hydro-Social Dilemma*,
https://doi.org/10.1007/978-3-032-08941-0_7

'best' means in this context, and to whom it is addressed, means something different to different groups of actors. The goal here is to map the different types of actors and the solutions they propose. This way, the controversy once more unfolds the manifold ontologies on water per se—and for Lake Urmia in particular.

7.1 Aral Sea Syndrome

The Aral Sea is a tragic landmark in modern human-water relations. Once 68,000 km^2 in size, the endorheic saline lake was targeted by the Soviet Union in the early twentieth century as a focal point for agricultural production, particularly cotton. As a result of the intensive use of the Aral Sea's tributaries, the lake split into two separate bodies of water, with the water level falling by 16 m in the northern part and 23 m in the southern part, reducing its surface area by 80% and the water volume from 1,093 km^3 to a mere 98.1 km^3 (Tussupova et al., 2020: 10). Once the fourth largest lake in the world, it has mutated into a desert called Aralkum (Opp et al., 2017: 171), and where there was once water full of life, only rusty fishing boats bear witness to its former 'glory'. These images of the Aral Sea are etched into the collective memory and carry a weight, a level of gravitas, that gives a certain presentiment of the Anthropocene—a prelude to what could happen to any river, lake or wetland if human, primarily economic, interests are put above all else. Words and images have agency, i.e., they act on any state of mind, insofar as one remembers that action does not require intention but merely an action carried from one actor to another—or perhaps not. Accordingly, the disappearance of the Aral Sea marks a starting point in global environmental history, where every subsequent disappearing body of water is always traced back to this first major event. It is almost as if there is a new era in the human-water relationship: the one before the Aral Sea disaster and after, the subsequent Aralkum, the man-made desert flanked by the shadows of a former lake.

In this regard, the photograph my colleague took at Lake Urmia (Fig. 7.1), emphasises relatedness: the picture may show Lake Urmia, or rather a salt desert, a rusty fishing boat and a glum woman, but it *reminds* us of the Aral Sea tragedy. Hence, the question arises as to whether the Aral Sea and Lake Urmia share a similar process and, as such, the same outcome. This is the main concern of local residents, scientists and politicians, namely that they will become victims of Lake Urmia's *mutation*—an unfamiliar and hostile environment.

Fig. 7.1 Lake Urmia during a field visit in Sep/Oct 2019. © Sebastian Transiskus

The process of mutating water bodies has led one group of scientists to identify the phenomenon as one of 16 syndromes representing global human–environment change. The Syndrome Approach applied by the German Advisory Council on Global Change uses the medical metaphor to indicate a particular *sickness*, a 'clinical picture […] [which is] a means for specifying an appropriate therapy for the patient' (Lüdeke et al., 2004: 42). The Aral Sea Syndrome was declared as 'environmental damage to natural landscapes as a result of large-scale projects' and a 'tendency towards top-down project planning and purely technological solutions' (Lüdeke et al., 2004: 43–44). The Syndrome Approach defines a particular set of reasons why a body of water desiccates' and these are entirely anthropogenic and, even more so, highly political, economic and technocratic. Nevertheless, as the former chapter elucidated, the reasons why Lake Urmia has desiccated are controversial. The prevailing wind, however, has shifted towards the anthropogenic impact, with AghaKouchak et al. (2015) declaring that the 'Aral Sea syndrome desiccates Lake Urmia'—with farmers and unsustainable irrigation practices being the pathogenic agents, while a lack of governance and action provide the momentum for these collective-threatening entities.

The 'catastrophic death of the Aral Sea' (AghaKouchak et al., 2015: 307) will be followed by Lake Urmia, or what has already been described as 'a striking, disastrous testimony of the syndrome (Schmidt, 2021: 190). This is the power attributed to the Aral Sea, which acts both as a reminder of what happened to such a large body of water and what could—and will happen—to other lakes, as if a disease is spreading across the globe, suffocating water collectives. In other words, this event stands at the very beginning of a new era, where slow onset events will dictate the future of water availability to humans and non-humans. As a reminder, however, the disease is caused by top-down mismanagement, which implies a lack of proper water governance, i.e., the narrow, mainly economic and ideologically overloaded approach to water causes waterbodies to go 'extinct'—or more realistically, 'leave' the local water collective in the long run. The backward-looking or retrospective responsibility (Doorn, 2019: 161) for the Aral Sea Syndrome stands squarely with central government. Moreover, as Tussupova et al. (2020: 16) point out, the conflict is not only based on a '"top-down" constellation, but is similarly characterised by the geographical and metaphorical distance between central decisionmakers and the Aral Sea collective or the victims in general; that is to say, as long as problems are spatially speaking kept at bay, they are "out of sight, out of heart"' (Tussupova et al., 2020: 16). The Aral Sea event, as much as the syndrome, has radiated out to Lake Urmia, its local residents, decision-makers and scientists. It is used both as a comparison to Lake Urmia and as a constant reminder that if it is 'business as usual', the outcome will be similar to the Aral Sea (Abbaspour et al., 2012: 257). As such, 'people are afraid of similar consequences' for Lake Urmia and its adjacent population (Rouw, 2017: 34).

As a recipient, however, one must inductively put the pieces together; in other words, what happens at Lake Urmia happened at the Aral Sea, and hence the causes—according to the authors—must be the same. The similarity they provide is that a large body of water desiccates due to 'over-exploitation of input water, [which] *may be* the *main driver*' (AghaKouchak et al., 2015: 310; emphasis added). This over-exploitation is indeed linked to the causeway, proliferating dams and irrigation projects, which indicate a 'managerial and socioeconomic myopia' (AghaKouchak et al., 2015: 311). However, at the same time, management is 'uncoordinated and disintegrated', while 'climate change adds additional pressure' (AghaKouchak et al., 2015: 310–311). Another article, written by Schmidt (2021: 190), correlates the Aral Sea Syndrome with a 'complex set of causes'. However, the original understanding of the syndrome is straight to the point: top-down mismanagement and large-scale projects—no more, no less.

The point here is not that the authors are wrong to use the term 'Aral Sea Syndrome' altogether but that they overload the term in its original sense, whereas several syndromes are at work at Lake Urmia. This is why the controversy over the causes has not yet been resolved: if the causes are so obvious, why do scientists disagree? Why is it the 'tyranny of small decisions' for Pouladi et al. (2021: 9), primarily droughts, secondarily anthropogenic impacts for Schulz et al. (2020) or the top-down totalitarian system for Karami (2018: 44)? Because 'like blind men describing an elephant', each study approaches Lake Urmia from a different angle or from a different starting point in the network, but they can never grasp the Lake Urmia collective as a whole, only parts of it, and in each part of the network there are different actors (made) responsible for the disaster. This being said, the inventors of the Syndrome Approach, the German Advisory Council of Global Change, emphasised two other syndromes in the Lake Urmia region: the Sahel and the Dustbowl Syndrome.[1] The former indicates 'overcultivation of marginal land', due to poverty and a lack of different sources of income other than farming, while the latter identifies the over-exploitation of soil and water in the face of capital-intensive agro-industries as a core issue (Lüdeke et al., 2004: 48). Yet, next to these syndromes, Madani et al. (2016) identified 17 drivers for the looming water crisis in Iran/Lake Urmia. The point here is that the list of drivers is too long to be covered entirely be either one of the three aforementioned Syndromes. This is an important point to recognise, as manifold issues demand similarly manifold solutions. Lake Urmia is not just one body of water, whose reproduction is best described by an abstract input–output model—it is the conglomeration of numerous networks. In other words, all factors have the potential to influence the water level or ecological health of Lake Urmia. The actions taken, whether intentional or otherwise, have a cascading effect on the lake. These actions range from the small farmer digging another illegal well to irrigate their water-intensive crops, such as apples, which are marketed as a 'fruit from heaven' (Karami, 2018: 53), to the central government's populist measures to maintain low water and energy prices (Madani et al., 2016:

[1] The Aral Sea Syndrome was not relevant at Lake Urmia according to the authors, which is, how-ever, due to the fact that river discharge and dams higher than 30 metres were used as a dataset to determine the affected regions. Moreover, the data was outdated, stemming from the late 1980s, when dam constructions were far from their peak in Iran and at Lake Urmia.

1001). Indeed, there is a 'complex set of causes', which in part seems to stem from a totalitarian system and its ideological concept, termed by Karami (2018) as 'Aralism'—emphasising the link between Lake Urmia and the political responsibility pronounced even more than by the Aral Sea Syndrome. Nevertheless, Lake Urmia and the Aral Sea may share similar processes and probably the same dystopia, but the differences between the two cases must be established; otherwise, any solutions are doomed to failure.

The Aral Sea and Lake Urmia may have many factors in common: totalitarian regimes, the expansion of arable land, water as a source of development, delayed awareness of the problem, the concept of water as an exploitable common pool resource and ideologically motivated, top-down mismanagement (Karami, 2018; Opp et al., 2017; Schmidt, 2021). Both lakes are embedded in and characterised by a hegemonic utilitarian technocratic culture, indicating an anthropocentric positionality—which is the case at Lake Urmia, at least in the early stages. The fresh water running down the rivers was perceived by authorities as being wasted, a missed economic opportunity, if they could only continue to feed the salty lake (Hashemi, 2012: 109). As such, water must be manipulated to serve human needs and human development, or at least economic development. However, the manipulation of water, which is supposed to indicate a purely state-driven agenda, is untenable in the case of Lake Urmia. Although Karami (2018), Opp et al. (2017) and Tourian et al. (2015) highlight the similarities between Lake Urmia and the Aral Sea, the two cases differ significantly, as we shall now discuss.

Starting in 1918, the Soviet Union diverted water from the two tributaries feeding the Aral Sea in order to increase cotton production, referred to as 'white gold', indicating the economic value of the resource. From the 1930s, the aim was to reduce dependence on cotton imports, which led to the construction of large-scale irrigation canals to expand farmland in the Ghare Ghom desert. Neither the international community nor scientists warned of the consequences of such an undertaking. However, this was due to the Soviet Union's isolation from the international community, which in the face of the totalitarian regime's propaganda led to an unwavering commitment to the necessity of the project. The demise of the Aral Sea was a slow process, with the first signs of its mutation appearing in the 1980s, and especially in the early 2000s. Scientists referred to the tragedy as an 'accident of nature', absolving the central government in Moscow of any blame, even though decision-makers knew about the ecological and collective damage to the Aral Sea before the mega-project was initiated. Not only did those in power foresee the tragic consequences, but they consciously accepted them, because the economic, and therefore ideological, added value of the mega-project was paramount; the 'death' of the Aral Sea was therefore perceived as necessary quasi-collateral damage for the greater good. By insisting on the inevitable end of the Aral Sea—as a natural rather than as an anthropogenic phenomenon—scientists made themselves jointly responsible for the disaster (Karami, 2018: 47–48).

The project, however, was an economic success story, while non-human and human local residents at the Aral Sea paid the externalities; that is to say, they had to live with the dire consequences in the form of dust and salt storms, loss of livelihoods (especially fishing), negative effects on the local climate, environmental degradation and loss of biodiversity (Micklin, 2006: 551–554). Nonetheless, tragedy for one side was success for the other. Even nowadays, Uzbekistan is the largest cotton exporter in the world, marking the mega-project as *economically* sustainable (Karami, 2018: 47).

In the case of Lake Urmia, though, I argue on the basis of my literature review that its mutation has occurred due to different factors.

First, the Lake Urmia basin is a semi-arid region with abundant arable land. As such, the region is considered the country's breadbasket. Therefore, the intended transformation to ensure food self-sufficiency was not as drastic as producing water-intensive cotton in the midst of a desert.

Second, in post-revolutionary Iran, the government's goal to secure self-sufficiency was guided by the ideas of autarky *and* the development of small and medium-sized farms, which were considered the socio-economic backbone of Iran (Shakoori, 2001: 113). The focus was on rural areas that had been neglected during the Pahlavi era. The project had, so to speak, a socio-cultural rather than an economic implication, since Iran's revenue is built on oil and not on agriculture (Darling, 2013: 208).

Third, the main difference between the two cases is the fact that the Iranian policy failed to reach its goal (Madani et al., 2016: 1000). On the contrary, Iran's food insecurity has increased since the Islamic Revolution (Shahi, 2019: 25–26). Even the excessive dam constructions during the Rafsanjani presidency in the late 1980s and early 1990s became 'an end in itself and not the means for tangible and sustainable economic growth' (Shahi, 2019: 27). Similarly, Hashemi (2012: 138) notes that

'Water resources planning have [sic] emphasised securing more water for agriculture but irrigation and drainage network constructions have lagged behind dam-building. There are many dams without even primary networks'. Despite heavy subsidies to boost agricultural expansion, these investments were arbitrary and unsustainable, and agriculture was thus left behind due to a lack of investments and less efficient water use (Madani et al., 2016: 1000; Shakoori, 2001: 167; 116). Conversely to the government's claim to improve livelihoods in rural areas, and as such the agricultural sector, urban areas and industrial sectors were favoured (Shakoori, 2001: 167; Madani, 2014). The government not only missed the target, but also co-produced Iran's water bankruptcy (Madani et al., 2016: 997).

Fourth, the Lake Urmia case proves how even an unsuccessful transformation—hence unintended (trans)mutation—can lead to such a detrimental event. If one imagines the agenda of the Soviet Union to transform the Aral Sea basin as an arrow, then this arrow would be a straight line of action from the politburos in Moscow to the Ghare Ghom desert, or cotton farmlands and indirectly to today's Aralkum. Put simply, it is easier to place retrospective responsibility on the government, because its implementation succeeded. In Iran, on the other hand, the government's failure triggered a cascade of irresponsible behaviour by numerous actors. Iran is marked by a lack of water governance, and hence a lack of groundwater regulations, competitiveness among provincial stakeholders over water resources, farmers' lack of participation in reducing water consumption and so on (Madani, 2014; Madani et al., 2016; Nabavi, 2018). Furthermore, because Iran's policy failed and the arrow of action was abbreviated—starting of the politburos in Tehran—other actors and their actions unfolded. The government was responsible for its inaction, but the technocratic and utilitarian paradigm still prevailed, albeit in a less straightforward way, as the tens of thousands of illegal wells, water diversion projects and illegal water withdrawals in the Lake Urmia basin demonstrate. Responsibility for the lake's desiccation, so to speak, is distributed among more actors—although not evenly, of course. Water management is myopic and disintegrated and not accurately planned from start to finish, and in itself it is incapable of mutating Lake Urmia 'single-handedly'. The mutation occurred due to a cascade of actions and not one single arrow of action, to which everything fell in line like in the Soviet Union. Accordingly, the comparison established between Lake Urmia and the Aral Sea, and as such the identification of the Aral Sea Syndrome as the driving factor, is in my estimate not correct but misleading, as responsibility was diversified among many other actors than only the central government and its lackeys. In short, a lack of top-down responsibility for water resources paved the way for the tyranny of small decisions, and Lake Urmia was consequently consumed by both 'top' and 'bottom' actions. The Soviet Union gained control over its water resources to the detriment of the Aral Sea, but the Islamic Republic of Iran never gained control over its resources, to the detriment of Lake Urmia—not implying that the urge to 'control' water would do any good by default.

Furthermore, and fifth, scientists have opposed the government's emphasis on natural factors as the main reason for Lake Urmia's desiccation, in contrast to scientists in the former Soviet Union in the case of the Aral Sea. As we have seen herein, although the main cause is a controversially debated topic, scientists have acknowledged the anthropogenic influence all along and not neglected it in favour of governmental propaganda. This opposition is even more openly articulated by Iranian academics working at foreign universities.

Sixth, and lastly, the desiccation of Lake Urmia is a comparably much 'faster' slow onset event, triggered by drought first and then followed up by water overuse by the agricultural sector (Alborzi et al., 2018; Schulz et al., 2020). The Aral Sea Syndrome, however, clearly demonstrates that the anthropogenic factor was the almost sole reason for the water body's desiccation. Yet, although the agricultural sector in the Aral Sea basin is responsible for the decline from 1992 to 2015 (Su et al., 2021), climate change seems to have gained the upper hand and is now considered by Huang et al. (2022) to be the main reason. The causes are, so to speak, somewhat reversed.

The mutation of the Aral Sea and Lake Urmia shares similar driving factors, but the differences—as can be seen—are much too drastic to refer to the Aral Sea Syndrome as the sole active 'disease' in the Lake Urmia basin. To stay with the medical metaphor, there are too many agencies involved in the desiccation of Lake Urmia, and thus too many syndromes at work at once. This leads to the conclusion that developing a roadmap to restore Lake Urmia will not suffice if the Aral Sea Syndrome is the only issue made responsible. To distil this notion, it is important to acknowledge a separation between the similar *outcomes* of both lakes and the *processes* leading to the disasters, which are not particularly similar. The Aral Sea and this argument will be better understood in the following chapters, referred to in the literature and especially in the ULRP as a *trauma*—a vivid warning sign that by default justifies any measures taken to restore Lake Urmia. Without the disaster at the Aral Sea, scientists, decision-makers, politicians, media and local residents would have to imagine a fictional dystopia. The Aral Sea, or rather Aralkum, *grounds*—literarily—this fear, as people are transported through their imagination to the very place they wish never to see replicated for their home. They may refer to other lakes as sister lakes (Great Salt Lake or Aral Sea), but they would not do so with a desert—a desert is nothing but a painful memory of water.

7.2 The Push and Pull of Utopias and Dystopias

Scientists, politicians and activists warn of an impending mutation—not a crisis, as Madani (2014) suggests, because a crisis would indicate that the problem can be circumvented or at least endured until it is over (Latour, 2017: 7). According to the discourse, the current trajectory of Lake Urmia points to the same dystopian outcome as what has befallen the Aral Sea. Therefore, the vexing question for the Lake Urmia collective is: How can we avoid the Aral Sea fate? However, before describing and historically embedding the various solutions, the central research question must be kept in mind: What is driving the hydro-social metamorphosis of Lake Urmia? Is it only the fear of a possible dystopia, or is there something else? Those who want to restore it are driven by both negative projections into the future *and* reminiscent memories of its former 'glory' or idealised state. Against this backdrop, the Aral Sea is used as a template to explain how the same negative processes are leading to Lake Urmia's demise—with salt being declared as the main antagonist of this new, dystopian landscape. Vast salt flats are in the process of displacing the last remnants of Lake Urmia, and due to wind erosion, they are spreading far beyond the boundaries of the catchment area. The environment, agriculture and rural and urban populations are threatened by these salt particles, which collectively harm the health of the region (Khatami & Berndtsson, 2013; Nhu et al., 2020). Fear is one driver, but there is another important one: time, or rather *urgency*. Studies have warned that if the negative factors affecting Lake Urmia continue, the lake is likely to disappear within a few years. Abbaspour et al. (2012: 257) predicted 10 years until the lake dries up completely; Tourian et al. (2015: 358) calculated the years around 2020 as being the last of Lake Urmia. Whether or not a time stamp or an ultimatum has been applied, the sense of urgency is omnipresent across the sciences (AghaKouchak et al., 2015; Haghighi et al., 2018).

To the dystopia, the time pressure, the already felt consequences of the salt winds is added another fear: the collapse of social life in the region, accompanied by mass migration. To exemplify this point, three statements with similar implications will be presented. In a speech at the UNPO in 2015, a member of the Southern Azerbaijan Student Movement argued that if Europe did not make an effort to save Lake Urmia, 15 million people would migrate to Europe—with all its consequences (UNPO, 2015; AZOH n.d.). Here, the threat was made quite explicit, used as a tool, again a dystopian one, to mark Lake Urmia as something that even Europe had to act on in order to avoid far-reaching consequences. Similarly, Rouhani (ISNA, 2018) used Lake Urmia as a political tool, a possible stabiliser or disruptor, when he prematurely stated that restoring it had saved 14 million local residents from mass migration, as well as adjacent provinces from mass immigration. Issa Kalantari, head of the ULRP and the DoE, clearly expressed his dystopian view, stating that Tabriz would have to be evacuated if the lake disappeared (Bakhtiari, 2020). However, there is another driving factor lurking behind the two aforementioned ones. Because Lake Urmia is interwoven extensively with numerous actors and actants, its mutation has literally extensive, negative consequences. Thus, its desiccation is not only an issue for adjacent local dwellers in the basin, but it is also perceived as a national and, occasionally, an international threat.

In this context, interviewees noted:

> '[Lake Urmia] is an international problem. Because natural resources such as water, lakes, etc. have non-use values (existence, option and bequest values), and these values are not related to the nations'. (Expert-2)

> 'Neither an international nor an Iranian problem, but a regional problem because of the flood of immigrants and possible dust storms stemming from that'. (Expert-3)

> 'In my opinion, the Urmia Lake problem is an international issue. Because the region's ecosystem is affected by it'. (Expert-4)

Most interview partners said that Lake Urmia is an international issue, the extent of which was expressed by the universal, non-economic value of water, the 'flood of immigrants' and dust storms and the region's affected ecosystem.

The Aral Sea as a form of 'fiction' has already been identified as a driving force in the Lake Urmia hydro-social collective. The dystopias that have just been presented, which are placed in relation to the Aral Sea—or comparable lakes—function both as a warning and as an obstacle to be avoided at all costs. However, next to the dystopia is another invisible actant that tries to shape Lake Urmia by its image. This 'idea' is for some a source to pull Lake Urmia's trajectory away from the dystopia, to redirect it towards a better future, reminiscent of the pre-crisis past. For others, this idea may be a source of psychological distress, as the following study argues.

In their case study at Lake Urmia, Ženko and Menga (2019) sought to establish a link between mental health and water scarcity. From their perspective, 'water has many dimensions, and people form their identities *around it*' (Ženko & Menga, 2019: 9–10; emphasis added). Moreover:

> […] because of farmers' strong attachment to their land, their occupation is intertwined with their identity, and so farmers often perceive the degradation of their home environment as a threat not only to their livelihoods but also to their self-identity. (Ženko & Menga, 2019: 9; emphasis added)

In other words, water (and engagement with it) creates a specific human–environment identity. Given this basic assumption, it follows that engagement with the environment also creates an identity that feels connected to that environment. Yet, at Lake Urmia, this identity has been *shaken* by environmental degradation and a lack of water. Farmers feel deprived of their profession, of their way of life, which was linked to the environment. Many farmers were categorised as showing symptoms of *solastalgia*. The term is used to describe people who feel homesick, even though they live in the same place, but can no longer recognise it due to drastic, negative changes. However, as the environment is an important part of their identity, especially in the case of farmers, this homesickness sometimes leads to severe psychological distress, which can go as far as depression. This is where the article by Ženko and Menga (2019: 9) ends, but it remains rather vague about the substance of the farmers' solastalgia or identity. If one approaches these two things from a Latourian direction, there may be a possibility of working out an entity that is more than just a mental illness that ultimately takes place only in the inner world of farmers.

I think there is a fundamental problem when, as Ženko and Menga (2019) write, 'people's identity is formed around water', because, as a consequence, this would mean that two insoluble entities would be merged, i.e., the subject's mind and the material object. Of course, the argument is that working by hand, spade and plough makes this connection a habit, one that can become part of the farmer him- or herself. Nevertheless, the question remains: What is this connection between a thing and a person? Subjects have an inner world, clearly separated from the material world full of objects. And what about identity? One solution to give an explanation to the complicated interactions between nature, humans and identity is by introducing Latour's (2013) idea on *fiction* in his book 'Modes of Existence'.

I have already touched on fiction in Chapter 6.1, where Latour used the metaphor of wolves roaming around, to express how these beings carry us away, i.e., they make someone imagine something at a different place and/or time. When fiction and reference cross—a tool used by sciences—these fictional beings report what happens on the ground through reliable references. However, fiction is also capable of crossing the material, such as Lake Urmia. In this case, fiction reveals itself to the observer through the aesthetics of the observed object. Plainly said, what the observer experiences as aesthetic sticks with him. As a result, Lake Urmia is no more just a plain lake full of salty water. The lake received a name, while the Azeri relate to it as the 'turquoise solitaire' (Mirchi et al., 2015), poems were written, pictures painted, fairytales told and memories kept—in the mind or in photographs (see e.g., Transiskus, 2021).[2] In this sense, there are two anchor points for Lake Urmia: one is the material lake, which is nevertheless fading, and the second is the imagined, fictional landscape. The former fails to exist if there is not enough water to fill the lake bed, while the latter goes extinct if no one remembers the turquoise solitaire. This being said, and to return to solastalgia and identity, it is this fictional Lake Urmia that both constitutes a person's subjectivity and his identity and is also capable of disrupting the latter.

[2] According to Latour, people do not create these works of art or this fictional landscape, they constitute them – '[the beings of fiction] impose themselves on us after imposing themselves on those responsible for their creation, for the latter are more like constituents than "creators"' (Latour, 2013: 240).

Certainly, not everyone's identity is shaped equally by this fictional landscape, and art, memories or even symbols (e.g., ULRP's logo) are subject to interpretation. Nevertheless, '[i]f the work needs a subjective interpretation, it is in a very special sense of the adjective: we are subject to it, or rather we win our *subjectivity* through it' (Latour, 2013: 241; emphasis added). To contrast Ženko and Menga's (2019) position that identity shapes around water, it is the fictional Lake Urmia that constitutes a person's identity, while its aesthetics, artworks and so on are its conduits, or vessels. In this sense, I argue that it is easier to convey the argument that humans and the environment (or nature) can form a bond through the invisible beings of fiction—that missing link, in my opinion. Subjects and objects are thus not separated, and humans receive parts of their identity from the outside, with fiction as the material used to weave the web of a person's subjectivity (Fig. 7.2).

This alternative representation of the human–environment relationship and identity differs from Ženko and Menga's (2019) position, insofar

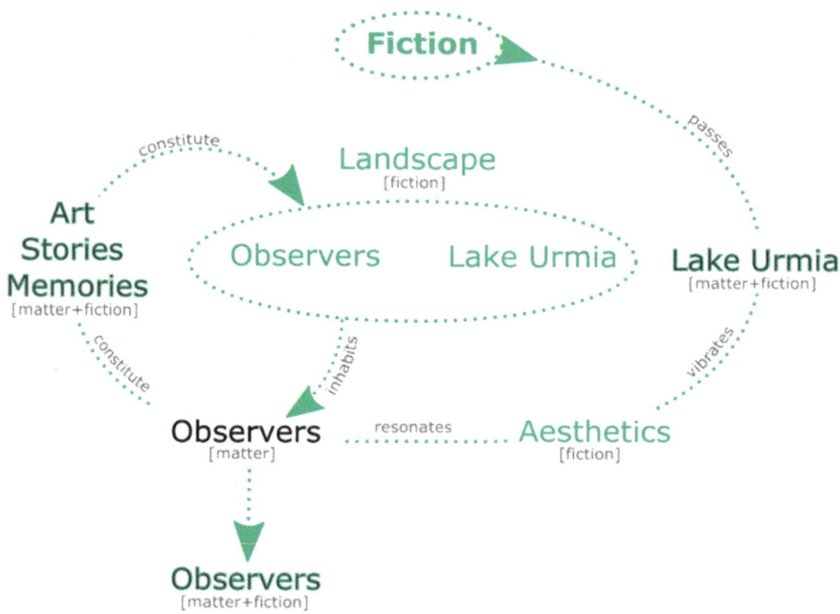

Fig. 7.2 Fiction and the process of fictional landscape as part of human-identity

as identity serves as a medium between the human and the environment. With this in mind, I offer a different angle from which to perceive a phenomenon such as solastalgia, since it may be somewhat easier to explain why change in the material environment has an effect on the human state of mind. There is now also a link established through the fictional landscape that feels real to people—or, nowadays, surreal in case of the new Lake Urmia. Although fiction still passes through this 'new' Lake Urmia, such a salt desert also exerts an immense attraction, it no longer resonates with the local population.

While tourists may be fascinated by such landscapes (see the Aral Sea in Tussupova et al., 2020), the altered landscape and associated environmental degradation create dissonance between local/regional observers and the 'new' aesthetic. The reason for this dissonance is the lack of correspondence between the fictional landscape of Lake Urmia, which is inherent in and part of people's identity, and the material landscape. When I visited Lake Urmia, I still perceived the salt desert, the new Lake Urmia as something aesthetic. However, my perception of what it had become was always challenged by the fictional landscape of the local population. Interviewees told me how vivid and populated Lake Urmia used to be and how close the shoreline was to their doorstep, whereas now it is miles away in the distance. The fiction of the former Lake Urmia is omnipresent and thus has overwritten my understanding and image of the lake; I can understand why this new desert-like landscape may be less aesthetic than I perceived, since now there lurks at the back of my head the fictional landscape created by the regional population.

When protesters in Tabriz and Urmia lamented the death of Lake Urmia at the hands of the Majlis in 2011, and the hashtag #IamLakeUrmia spread across social media in all the years since, they were speaking the truth in both cases, in that the virtual Lake Urmia and its landscape are a part of them that is in danger of disappearing, because the new, material Lake Urmia no longer 'vibrates', no longer emits the old familiar resonance, while the dissonance it has created shakes the inner world of its inhabitants (Fig. 7.3). This shaking is precisely the solastalgia identified by Ženko and Menga (2019), i.e., the feeling of homesickness despite being in the same place. The phenomenon occurs when there is a break of planes between the fictional landscape and the new, material one—still passed by fiction but vibrating differently. The discrepancy between these two planes causes dissonance, which may be imagined like transforming tectonic plates. The longer the difference between the two planes prevails and the larger it becomes, the stronger the dissonance grows that 'sweeps' over observers and as such threatens their identity. This may take place in a staggered manner; that is to say, people may accept changes to the real landscape and adapt to them, but accepting dissonance between the material and one's fictional landscape might be hampered by an enduring *psychological* crisis. In this sense, solastalgia (dissonance) is an entity that might be similar to what Latour (2013) referred to as 'metamorphosis', *psychotropes*. Latour (2013: 192) elaborates on this point as follows: 'As if there were many beings that have to be called psychotropes in the sense that they modify who you are almost completely; they get you all mixed up by making your soul spin'.

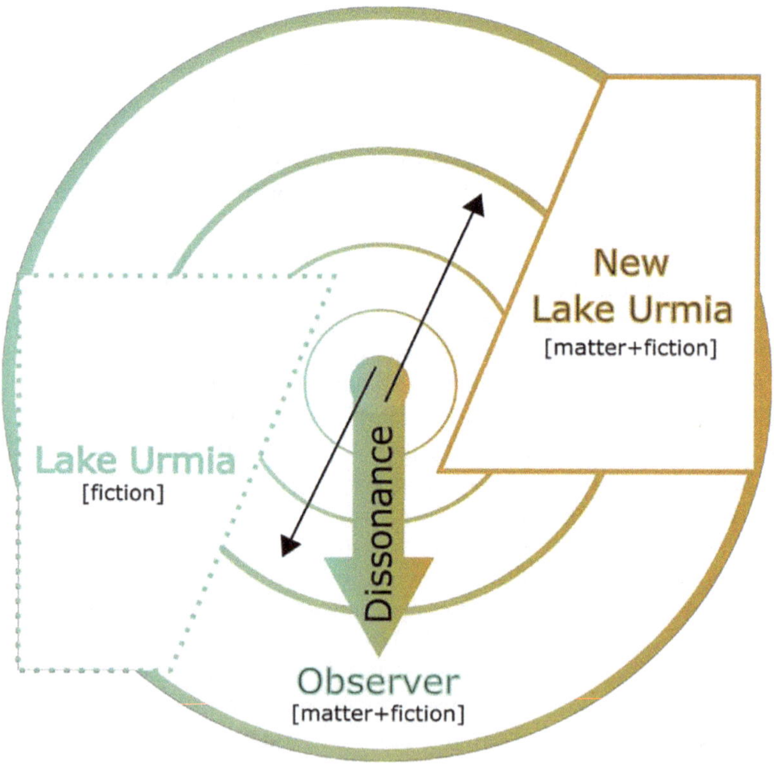

Fig. 7.3 Breakage of identity (fiction vs matter)

If one accepts this rather radical perspective on the relationship between the material Lake Urmia, the fictional Lake Urmia and people's identity, then a completely new field could open up, because now, it might be tempting to investigate the different inner, virtual landscapes. These have been touched on, albeit only superficially and usually dismissed as misconceptions.

Pouladi et al. (2021), for example, shed some light on this fictional world by conducting research on local farmers. Even if the researchers dismiss their findings as merely the perspectives or perceptions of farmers, the fictional landscape can still be pieced together. The 'memories of the past', as Pouladi et al. (2021: 8) call them, passed on by farmers' ancestors, allow a small glimpse into this world (or worlds) that is—for obvious reasons—inaccessible to the natural sciences. The landscape of the fictional Lake Urmia is characterised by two features that have a lasting effect on the actions of farmers, at least those interviewed. First, from their point of view, there is an abundance of water in this landscape, i.e., the 'normal state' is an abundance of water. Second, the landscape is not static but dynamic and changeable, based on local experience. Droughts, and therefore the decline of Lake Urmia, are a cyclical phenomenon (as the scientific community would probably agree). This cycle is therefore an inherent aspect of the fictional landscape, so that over time, the lake returns to its water-rich state in response to a decrease. Precisely for these reasons, and thus also somewhat in contrast to Ženkos and Menga's (2019) solastalgia, there is no need for farmers who are convinced by these concepts of the fictional landscape to take action—sooner or later, everything will turn out for the better anyway.

Moreover, these fictional landscapes are even governed by different laws. Take, for example, people's religious framing of the causeway in Chapter 5.3, the unity of nature that should not be interfered with by anthropogenic factors. Or, on a different occasion, when protesters used the illustration of an anthropomorphised, grief-stricken 'Lady' Orumiyeh to show their dissatisfaction with the government's inaction (Ranjbar, 2022: 8). These examples might also show that the fictional landscapes of Lake Urmia contradict the material one. Although one could dismiss Lady Orumiyeh as a mere symbol or a reflection of the protester's mind, it might nevertheless be possible that this fictional character is one of the many beings in these fictional landscapes that contrast with the material one.

7.3 Overview of Solutions According to the Sciences and Experts

There are plenty of reasons to save Lake Urmia. From a material perspective, the lake would provide numerous eco-services, ranging from stabilising the local climate to biodiversity. Nevertheless, as I have argued above, these material factors might not be the only ones that play a role in hydro-social dynamics, since there are also immaterial factors that drive people and, as such, decision-makers. On the one hand is the dystopia, the fear of what is to come, as the current trajectory of Lake Urmia is pointing towards a 'dead-end', the becoming of another twin-sister lake of the Aral Sea. On the other lies the fictional landscape(s), which resembles a pre-crisis state that is to be restored. Yet, one could even say that the wish to 'restore' Lake Urmia is actually ***inspired*** by the past or, according to the perspective elaborated above, the fictional landscape(s). Put differently, these two drivers (utopia and dystopia) are opposites, like the relation between a protagonist and an antagonist, or a thesis and an antithesis. They are in some sense actors who have an influence on the solutions chosen by stakeholders. The issue, as will be shown, is that these stakeholders do not agree on how to resolve the tension between the two poles or, in a simpler way, how to restore Lake Urmia with a certain agreed upon goal in mind.

While some controversies are discussed only by experts (*A. urmiana*; ecological water level), the solution controversy enters public life at full blast. In order to get a better handle on this topic, four different perspectives on the matter will be elucidated in the following order: the sciences, central government, the ULRP and farmers. The important categories that will be of interest here are how the actors wish or recommend to solve the issue, and which values (and therefore ethics) they insist on. While the different ethical positions have already been mentioned and worked with (anthropocentrism, ecocentrism, partnership ethics), the two main categories in terms of solving water scarcity are supply- or demand-oriented strategies (Groenfeldt, 2019: 18), based on an economic view of water scarcity. Therefore, when there is not enough water to meet demand, there are usually two options: either reduce water consumption (demand-focused solution) or add more water to the hydro-social collective through predominantly technical solutions (supply-focused solution). However, the decision-making process is described as a 'moral choice' (Groenfeldt, 2019: 18), and as such every solution and every proposed technology is guided by morality—and politics. Politics, again in the Latourian sense, means that certain actors, human and non-human, are either accused of being responsible for worsening the situation, now and in the future, or other actors are presented as solution to the crisis. Who has to leave the circle of the collective, and who has to enter it in order to stabilise hydro-social relations for the current iteration?

As Fig. 7.4 shows, the vast majority of scientists and experts suggest demand-focused solutions. The reason why I made a distinction between direct and indirect measures is that while the former aims to reduce water usage by imposing restrictions or implementing new, efficient technologies, the latter aims at changing the context, for example by enhancing people's livelihoods or improving water management. Furthermore, another distinction between the two categories lies in their diverging ethical stances. In the case of the direct demand-focused solution, nature, or the lake, is the centre of attention (ecocentrism), while indirect measures aim to improve both the environment and socio-economic and, occasionally, political factors (partnership ethics). Conversely, supply-focused solutions promote anthropocentrism, since their primary objective is to increase water availability to improve the local or regional economy. The first pie chart in Fig. 7.4 shows how often these solution categories were mentioned, while the second pie chart shows which solution categories were mentioned per scientist or expert. The difference is important because it shows that different scientists and experts sometimes suggest different combinations of solutions categories. Nevertheless, what has become evident is that only a minority of scientists suggest supply-focused solutions, and even more so only two papers have suggested them exclusively (Golabian, 2011; Zoljoodi & Didevarasl, 2014). Based on this first, rough estimate of how scientists and experts are distributed in the solutions controversy, Fig. 7.5 elucidates the exact measures that have been proposed.

Demand-focused solutions primarily target the agricultural sector. The main goal is to transform agriculture, or as Madani (2014: 320) demands, 'Iran seriously needs its agriculture to be modernized/industrialized and become economically efficient'. I will return later to the controversy surrounding this view. In particular, it is believed that especially by decreasing irrigation water demand by increasing its efficiency, e.g., through drip irrigation, 'the dual aim to satisfy the increasing need for crops, and a limitation on water resource withdrawal', is accomplishable (Shirmohammadi et al., 2020: 8). Ethically speaking, the goal to

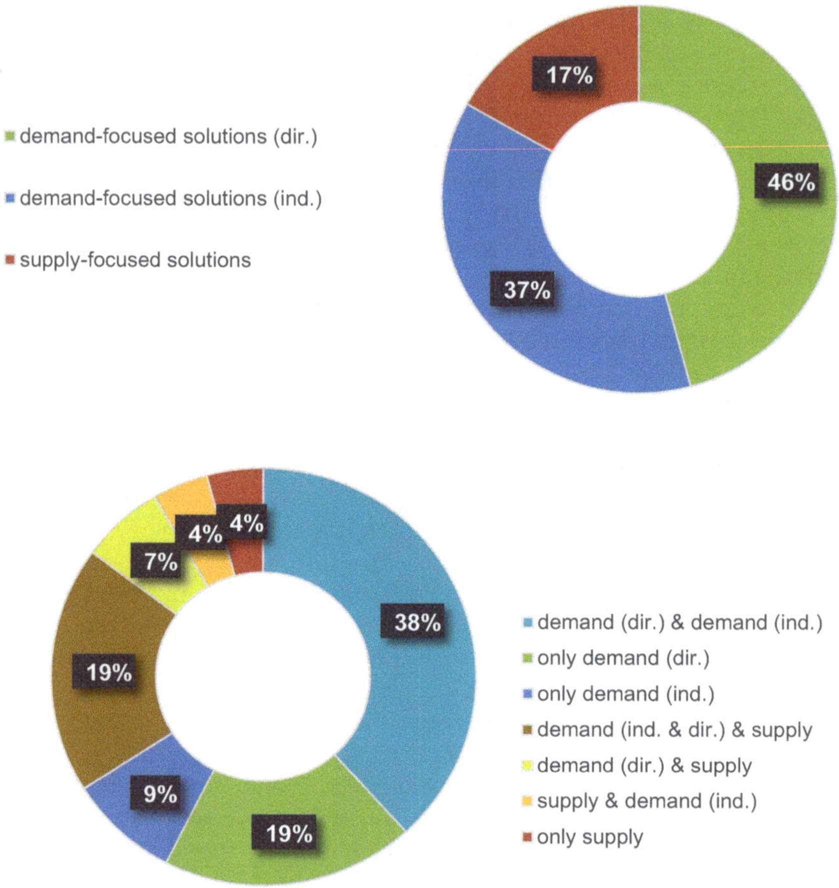

Fig. 7.4 Solution categories mentioned by scientists and experts ($n = 46$)

're-establish a balance between the health of Lake Urmia and economic development' (AghaKouchak et al., 2015: 311) is marked by supporting both the farmers and Lake Urmia. As such, increasing irrigation efficiency through technology could be characterised as a partnership ethic. To stay with the 'balance' metaphor, since modern technology is often considered to be a jack-of-all-trades, capable of addressing issues on both the ecological and economic fronts. The aim is to transform the system

Fig. 7.5 Weighted sunburst diagram of the solutions and categories proposed by scientists and experts

(demand- and supply-side) in such a way that the current crisis becomes a win–win situation.

Conversely, other solutions for transforming the agricultural sector are characterised by an ecocentric ethic. These proposals include changing cropping patterns and growing less water-intensive crops (e.g., from apples to grapes), stopping the exploitation of groundwater and reducing the amount of irrigated land. The main objective is to 'rebalance' in a similar way, but in this case by reducing water demand to relieve pressure on Lake Urmia. However, farmers are expected to bear the risks and negative consequences associated with these solutions.

In addition to changing material practices, six articles suggested focusing on farmers' perceptions, or rather educating them. The lack of awareness and sense of responsibility for the Lake Urmia crisis is a theme that runs throughout the discourse. Salimi et al. (2019: 631), for example, lament a 'gap in public awareness[...] [and] perhaps due to poor economic conditions, they lack a long-term perspective'. In particular, farmers, who have been identified by most scientists and institutions as the main responsible actors, should 'enhance their agricultural knowledge and awareness of environmental processes in the Lake Urmia basin' (Pouladi et al., 2021: 8). Moreover, farmers who do not seem to feel responsible for the demise of Lake Urmia (Farahmand et al., 2023: 13) must have a change of mind. In other words, from a hydro-social perspective, they must change their material practice as well as substitute their 'water concept' with the scientific one, in order to 'see things as they are' and begin a self-reflective journey that ends with taking responsibility for their unsustainable actions.

These solutions imply the transformation of the agricultural sector from head to toe. Politically, certain irrigation practices (e.g., surface irrigation), water-intensive crops, illegal wells and archaic perceptions of the hydrologic landscape of Lake Urmia and its basin must leave the circle of the hydro-social collective. This, according to the experts and scientists, would enable the collective to 'live together' in a sustainable way.

In addition to transforming the agricultural sector, experts and scientists see the improvement in water management and the implementation of water governance in the basin as having the potential to overcome the crisis, or at least to transform the human-water relationship from an unsustainable to a sustainable one. For example, Madani (2014: 322) argues in favour of changing the 'current Iranian water management paradigm [...] [to] prevent the development of similar problems in the future and secure sustainable water resources for Iran'. Water management should turn from its myopic, disintegrated state to one guided by Integrated Water Resource Management (UNEP 2012: 136), while regional water management should be institutionalised, turn from a reactive to a proactive state of mind and reorganise the 'current water governance structure and [empower] the Department of Environment' (Madani et al., 2016: 1009). Other solutions call for a more complex water management system in the basin that must incorporate 'social, institutional, and technical infrastructure' (Moghaddasi et al., 2017: 449). Moreover, different organisations, ministries and joint collaborations should increase and foster their co-operation (Salimi et al., 2019: 632). It is clear that water management is often related to the transformation of the agricultural sector, meaning that the latter only succeeds by implementing the former. Although water management should also aim at reducing water demand in the basin in general (Hosseini-Moghari et al., 2020: 1954), most scientists and experts correlate an improvement in this regard with the successful transformation of the agricultural sector. Water management as an alternative solution is mostly supposed to decrease water demand, but it seeks to mediate between the economic, social and ecological aspects and as such is meant to deliver a framework to improve the situation in the basin for the collective. In this sense, it is an alternative solution coupled with the desire to provide demand-focused solutions.

In third place, economic solutions should also reduce water demand or provide a cushion for farmers to ride out the transitional phase from traditional and unsustainable to modern and sustainable agriculture. In this sense, five authors explicitly mentioned that farmers should receive financial support in order to introduce new crops or overcome harsh drought periods in the future (AghaKouchak et al., 2015: 311; Moghaddasi et al., 2017: 449). One interviewee (Expert-2) argued that in the context of reducing water demand in the agricultural sector, and 'cultivating crops with lower water requirement [...] [a] real water price for irrigation water' has to be determined, which, additionally, calls for the establishment of water markets (Madani, 2014: 323; Saemian et al., 2020). However, one of the economic incentives has a controversial moral implication. Pouladi et al. (2020: 11) demand 'incentive mechanisms [...] to encourage farmers to be effective monitors, e.g., penalising rule violators (those who have illegal wells or extract more water than allocated rights) and rewarding monitors who catch rule violators by directly paying them a portion of collected fines'. Farmers shall therefore become the extended arm of the law, albeit social cohesion may suffer a significant setback, as they may begin to distrust each other and look for illegal wells, which almost everyone had during my time in the field. Therefore, not only could the lack of water 'dissolute' the social cohesion of rural communities (Ženko, 2020: 101), but similarly, this proposed 'incentive mechanism' will put increasing pressure on the social nature of these communities.

These economic strategies, whether morally questionable or not, aim at partnership ethics in the long run but are ecocentric in their immediate effect. Water will be re-evaluated and turned from a cheap common pool resource into a valuable good for farmers, which they have to organise and use proactively. In addition, they will benefit from these economic measures—receiving financial support, changing crops, decreasing their water demand—after the transition of the agricultural sector has been successful. The next economic solutions, however, propose a different direction, namely focusing on alternative sources for economic income in the region.

While the decline of Lake Urmia threatens especially the agricultural sector, the salt industry could emerge as a winner from the crisis. Although 400,000 tonnes of sodium carbonate ($NaCO_3$) are extracted annually, especially in Maragheh (southeast of Lake Urmia), and 50,000 tonnes by 'small private operators' (Eimanifar & Mohebbi, 2007: 4), other scientists see an opportunity to expand this industry. This would introduce 'a new source of income' (Khatami & Berndtsson, 2013: 8) while the high salt content of Lake Urmia 'could be used in the petrochemical industries' (Karbassi et al., 2010: 421). However, Karbassi et al. (2010: 423) suggest not using desalination plants, due to high economic costs, the emission of greenhouse gases and the adverse effects of brine disposal to the aquatic environment and its inhabitants. Conversely, Mardi et al. (2018: 49) are strictly against expanding the salt industry. Their findings suggest that disturbing the newly exposed lake bed to harvest salt would increase the amount of salt and dust distributed by the winds, which in turn would increase air pollution and hence increase the negative health impacts already prevalent in the region. The improvement in the industrial sector, in this case the salt industry, goes hand in hand with Madani et al. and's (2016: 1013) proposition to create 'alternative job opportunities' for farmers—a strategy familiar from the White Revolution.

Following economic, mainly demand-focused solutions, supply-focused solutions are intensely and controversially discussed in the sciences. As mentioned before, only a minority of researchers and experts suggest supply-focused solutions, while the majority turn adversely against these plans and propositions. The reason for this adversity is that supplying the basin with more water instead of using available water resources more efficiently and sustainably violates the core principles of demand-focused scientists and experts. For Iranian scientists situated in foreign, especially Anglo-American, universities, supply-oriented water management is the core issue and the main reason for the anthropogenic drought (Alborzi et al., 2018: 2). The hydraulic mission that is still active in Iran has to be stopped. As such, supply-focused solutions stem from the same 'Aral Sea Syndrome', which would continue to use water in an unsustainable manner (Madani et al., 2016: 1008). These 'quick fixes' (AghakaKouchak, 2015: 310) would backfire in the long run because in contrast to demand-focused solutions, they would maintain and perhaps intensify a water concept of abundance and ubiquity—and this concept is incoherent with sustainable, eco-friendly water management. To clarify, supply-focused solutions wish to maintain the status quo between water and humans—a nature control-oriented paradigm (Madani, 2014: 323)—whereas demand-focused solutions demand redistribution and rethinking the hydro-social collective—which leaves the impression that in order to *manage* water, institutions should be empowered to *control people*, albeit in this regard in the positive sense of the word. That the majority of experts are strong proponents of demand-focused solutions, and hence critical of supply-focused solutions, is also validated by Shadkam et al. (2017: 97). However, a closer look at these solutions reveals more than just a frontier between supply- and demand-driven resolutions. Two main solutions are offered by this other end of the spectrum: water transfers and cloud seeding. The latter is mostly perceived as a supportive factor that would help revitalise the lake in a 'new and effective way' (Barhagh et al., 2021: 24) by increasing local precipitation, thereby reducing the impact on droughts (Khatami & Berndtsson, 2013: 7). According to Zarghami and AmiRahmani (2017: 323), cloud seeding could increase precipitation by 7% and thus contribute 28% to the restoration of Lake Urmia, second only to increasing water efficiency. Cloud seeding is presented as both economically and environmentally sustainable (Khatami & Berndtsson, 2013: 7) and as an immediate solution to the crisis (Zoljoodi & Didevarasl, 2014: 358). However, other authors highlight the drawbacks of this technology. It is described as unreliable during dry periods (Golabian, 2011: 381), potentially ineffective (Azarnivand et al., 2015: 23) and controversial (UNEP & GEAS, 2012: 136). The spectrum ranges from those who see cloud seeding as a real opportunity to revitalise Lake Urmia, to those who see it as complementary, to those who generally distrust it.

In contrast to cloud seeding, the highly controversial topic of water transfers will be at first described only from its proponents' point of view. In the following subchapters (7.4 & 7.5), I will describe on what basis of argument actors disagree on the matter. Geographically, water transfers can be implemented in two ways: intra-basin and inter-basin (Khalyani et al., 2014: 6). Intra-basin water transfer aims to adjust the flow of water *within* the basin towards Lake Urmia in a more efficient way. For example, to reduce evaporation losses, the Zarrineh Rud, the largest tributary of Lake Urmia, would be transferred to the neighbouring Simineh Rud. Inter-basin water transfers, on the other hand, aim to use water from other basins. In Lake Urmia's case, the Aras River, flowing from Turkey along the Iran-Azerbaijan border, and the little Zab River, which is located south of the Lake Urmia basin and flows into Iraq, are regularly mentioned in this context (Khalyani et al., 2014: 6; UNEP GEAS, 2012: 135). Additionally to these neighbouring, transboundary basins, the Caspian Sea has been proposed by Golabian (2011) as a sustainable solution. However, in this context, the word 'sustainability' must be read with caution. Hossein Golabian, who worked at the time of his publication at the Center of Tehran University Architects, proposed to use the Caspian Sea, Lake Urmia's 'elder sister,' in his own words (Golabian, 2011: 382), to achieve a *permanent* and economically worthwhile solution. His core idea was to define an 'ecological water level', a 'utopian level' that was based on *A. urmiana*'s 'optimal' reproductive performance for three to four years during the mid-1990s to push an industrial production of *Artemia*. In other words, his idealised ecological water level set at ca 1,278 m is based on an economically lucrative idea of Lake Urmia—as an environmental but primarily economic asset. Hence, Golabian's water level is roughly 3 metres above Lake Urmia's long-term average of 1,275.35 m and 4 metres above the minimum ecological water level of 1,274.1 m (Zoljoodi & Didevarasl, 2014: 366). Golabian (2011) proposes to push Lake Urmia to this water level by using water from the Caspian Sea in order to enjoy primarily anthropocentric, utilitarian benefits. Among these benefits, he mentions the establishment of a large-scale *A. urmiana* industry and other industries—probably heavy industries—in need of large quantities of water, investment in aquaculture industries, installation of a new power source (osmotic electrical power plants) and the creation of new jobs, just to name a few advantages accompanying the Caspian Sea water transfer project (Golabian, 2011: 393–396). In addition, the project would include the installation of an overall 240 km-long pipeline, with the first section of the pipeline pumping water from the Caspian Sea up to 2,200 m. The pipeline would pass north of Tabriz, providing additional water for the city (Golabian,

2011: 386). Golabian (2011: 381) emphasises that a macro-engineering solution is necessary to prove enough water for Lake Urmia, and it needs more water than the minimum ecological water flow of 3,100 million m^3 suggests, because this amount would not cope with 'a hydrologic budget deficit'—or drought, to use a less economic term—as was the case from 2000 to 2003 (Golabian, 2011: 382). Ideally, the two best-case scenarios provide between 4,300 million m^3 and 6,750 million m^3 water to Lake Urmia, which would reach the 'ecological' water level of 1,278 m in 5.5 and 3.5 years, respectively. The cost of this ambitious project ranges from 4 to 5.5 billion US$ (Golabian, 2011: 393). Evidently, the Caspian water transfer project has to be marked as an ethically anthropocentric and highly technocratic solution, which revolves around water as a source of economic development. Water turns into a ubiquitous and an inexhaustible resource, since the Capsian Sea—and its volume of 78,200 km^3—provides all the water needed (Golabian, 2011: 382). High economic costs and uncertainty surrounding the ecological outcome of the project, however, make it a highly doubtful and, in the academic community, highly controversial solution.

Following the supply-focused solution, eleven experts and researchers emphasised the need to calculate and define environmental water flow. However, because this topic has already been discussed in Chapters 5.1 and 5.2, it will not be explained further herein.

Other solutions included democratising water management by involving more stakeholders—especially local people and farmers. Stopping dam construction in the basin and opening them up to release more water into Lake Urmia was mentioned overall six times. Lastly, social solutions received little attention, with four mentions and three to empower farmers.

To locate the researchers and their solutions, I have created Fig. 7.6, to visualise the researchers and their different positions on the matter. The main point of this map is not to represent the controversy accurately but rather to emphasise the distribution of experts and the clear tendency of scientists to choose demand-based solutions (directly or indirectly affecting water consumers' behavior) above supply-focused ones. With this in mind, the solutions proposed by other institutions can be compared to see how decision-makers hope to solve the crisis.

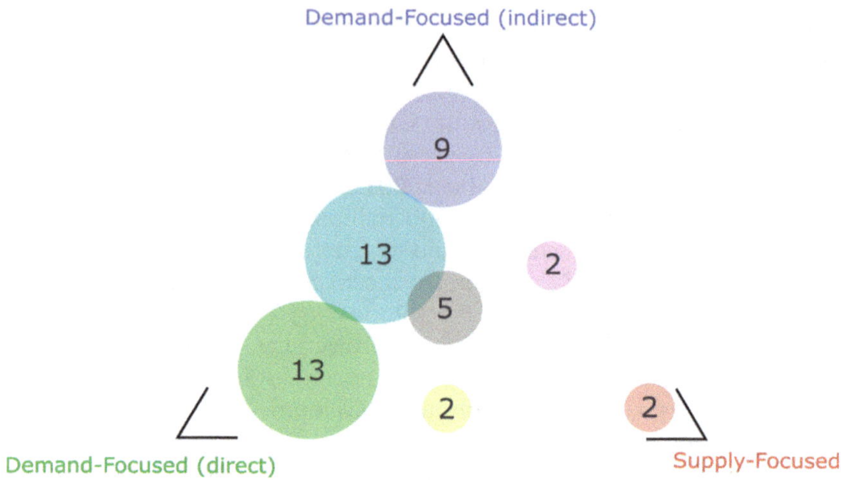

Fig. 7.6 Localisation of clusters and their predominantly proposed solutions

7.4 Lake Urmia's Restoration: First Steps and Controversial Debates

In 2008, the Department of Environment, the Ministry of Energy, the Ministry of Agriculture and the provincial governors of West and East Azerbaijan and Kurdistan formulated a Memorandum of Understanding (MoU), which was ratified in 2010 (Salimi et al., 2019: 614). The MoU consisted of a vision for Lake Urmia, a management goal, three main objectives guided by several principles, which will be elucidated in the following—and later on compared to the ULRP.

The MoU formulated a 25-year vision for Lake Urmia as follows:

> Lake Urmia will receive adequate water to sustain an attractive landscape and rich biodiversity where people and local communities can make wise use of its resources and will enhance co-operation between the involved provincial organisations. (MoU, 2010: 34)

The vision indicated how the lake, its environment and ecology, as well as people would be involved in this vision. Besides, it mentioned 'sustain[ing] an *attractive* landscape', which resonated with the fictional landscape of Lake Urmia elaborated above—as its aesthetics inhabited people's identity and demands to resonate with the material Lake Urmia. People were attributed with enough 'wisdom' or competence to make 'good' use of Lake Urmia's resources, which would 'enhance co-operation' between the three provinces. The vision for Lake Urmia in 2035 painted a picture of harmonious coexistence between the lake, its basin and its non-human and human habitants. To reach this vision, the MoU formulated an overarching goal, or core principle:

> To establish an ecosystem based management for the lake and its satellite wetlands within the context of sustainable development with effective involvement of all stakeholders including local communities. (MoU, 2010: 35)

The rather ecocentric goal set up by the participants was also reflected in their three main objectives: raise public awareness and encourage public participation, use land and water more sustainably and conserve biodiversity and make sustainable use of wetland resources (MoU, 2010: 36–38). The first objective, public awareness, included numerous PR campaigns under the lead of the Department of Environment (DoE) to ultimately make people and political decision-makers care for Lake Urmia, which should change 'their behaviour, actions and words' in the long term—the latter being related to politicians. One of the many actions was to 'organize mass awareness-raising activities, such as a human ring around the lake during the President's visit!!!' (MoU, 2010: 39). Furthermore, the objective aimed to implement livelihood strategies in wetland communities and to use wetland ecotourism as an additional source of public awareness. These changes could be characterised as 'soft-ecocentric' solutions, as they sought to indirectly bring positive change to the Lake Urmia basin. The main organisation responsible, and as such leading this first objective, was mainly the DoE.

In contrast, water authorities, the Ministry of Agriculture and lastly the DoE would lead the second objective, which dealt with water and land use management. This goal aimed to increase and maintain Lake Urmia's water level at 1,274.1 m—herein already considered as the ecological water level (MoU, 2010: 24)—thereby achieving good water quality in accordance with national thresholds, increasing irrigation efficiency in the long term by 15% and instigating land use regulation tailored to wetlands. In order to provide a sufficient amount of water to Lake Urmia, the MoU listed several actions, the following of which are of interest. While the solutions were primarily demand-driven, such as the enforced stagnation of agricultural water use, it was also investigated whether water transfers from neighbouring basins were economically and environmentally viable. Alternative solutions, such as reducing 'excess evaporation from the Lake', were also considered (MoU, 2010: 43). The MoU planned to reduce agricultural water use by introducing efficient technology, such as drip-and-sprinkle irrigation, fewer water-demanding crops, reasonable water prices and 'volumetric water delivery to farms' (MoU, 2010: 45). Moreover, illegal water extraction of ground and surface water was also mentioned.

The third objective was entirely dedicated to the conservation and revival of the lake's ecology—and thus the main responsibility of the DoE. Specific attention was paid to wetlands, flamingos, white pelicans, yellow deer, Armenian sheep, *A. urmiana* and the biodiversity of Lake Urmia's tributaries. As Table 7.1 shows, the MoU (2010) aimed, for some issues, at specific numbers to meet the third objective in the short and long term:

Overall, the MoU (2010) was mainly concerned with revitalising Lake Urmia and its ecosystem. In this regard, the ecosystem—and therefore ecocentric—approach was somewhat paradoxical, because the MoU

Table 7.1 Third objective of the MoU, and specific targets derived from MoU, 2010

Priority issues	*Short-term goals*	*Long-term goals*
Important Wetland Satellites	Increase area to 2,500 ha	Increase area to 25,000 ha or double the areas of southern wetlands
Breeding Population of Flamingo	1,000 pairs	4,000 pairs
Breeding Population of White Pelicans	750 pairs	1,000 pairs
Population of Yellow Deer & Armenian Sheep	Maintain present population + determine capacity of natural rangelands	Maintain present population + population living on natural rangelands
Population of *Artemia urmiana*	25 cysts per litre	40 cysts per litre
Biodiversity in River Systems in the Lake Urmia Basin	Two important rivers	All rivers

(2010: 36) stated that 'people should be at the heart of setting those objectives'. Even more so, traditional and local knowledge should be incorporated into water management next to evidence-based, and thus scientific, knowledge (for example in traditional *Artemia* harvesting or local experts and their 'excellent knowledge of local conditions' (MoU, 2010: 68)). These considerations, next to the inclusion of as many stakeholders as possible (29 stakeholders enlisted in the MoU's Appendix) and decentralised water management, indicate a primarily but not exclusively bottom-up process that would be guided but not entirely governed by powerful organisations. The MoU (2010: 60–64) listed a wide range of stakeholders, including—among others—farmers, reed, *Artemia* and salt harvesters, local communities, fishermen, buffalo herders, hunters and Islamic councils. Thus, Lake Urmia's management would be a mutual co-operation between local stakeholders and provincial/national organisations and would only be as successful as the sum of its smallest parts. However, the MoU (2010) was never able to realise its projects and thus achieve its 25-year vision, due to insufficient funding allocations (Salimi et al., 2019: 614).

The failure to implement the MoU allowed the disaster at Urmia Lake to continue unhindered. As mentioned above, there were protests and general unrest among the population in 2011. By the time President Rouhani came to power, the revitalisation of the lake was high on the political agenda as a symbolic and exemplary response to the country's environmental problems. Accordingly, in 2013, renewed attempts were made to reach Urmia's and its legally established ecological subsistence level. However, the first meetings between experts also revealed diametrically opposed views on how to deal with the crisis. For example, Salimi et al. (2019: 615) report that in the initial discussions, four separate groups pushed four separate approaches.

By far the most extreme proposal was to deliberately let Lake Urmia dry up and simply accept its fate. This perspective was led by Prof. Dr. Kardovani, a renowned desert expert (Kardovani n.d.). Kardovani declared Lake Urmia to be already dead and even dangerous to public life due to its polluting salt winds. Therefore, he regarded any attempt to save it as a 'betrayal of agriculture and the environment'. Revitalisation, he said, was illusory and only for show, because sooner or later the lake would be lost. In his view, all solutions, both supply- and demand-driven, were doomed to failure. Moreover, Kardovani was generally opposed to water transfers and cloud seeding, which he considers politically sensitive, technically difficult to implement and ultimately ineffective (Kardovani n.d.). Since Lake Urmia could not be revived, he perceived the only solution as limiting damage to people and the environment as much as possible, by removing the salt from the lake, greening the surface—to mitigate aeolian erosion—and eventually establishing a national park. Kardovani's principle was not to save Lake Urmia at any cost, as this would harm both the environment and agriculture—and thus large sections of the local population. In this context, it can be said that Kardovani considered Lake Urmia to be no longer part of the hydrosocial collective and was ready to abandon it completely. Although this position suggests an exclusively anthropocentric attitude, it should not be forgotten that his solutions also sought to ensure the continuity of the environment, albeit in a weakened form. Nevertheless, he placed farmers on a certain path dependency, as no alternative employment opportunities were available to them. For Kardovani, the preservation of agriculture was of paramount importance, and if a choice had to be made between the lake and agriculture, he believed that the latter should be preferred (Hamshari Online, 2012). This perspective is not shared by any of the analysed papers, and thus it has to be classed as an extreme, anti-Lake Urmia measure in the solution controversy.

A second group, led by Agh—*A. urmiana* expert from the University of Urmia—proposed to reduce the lake area artificially by constructing walls around a designated area. In his line of argumentation, the lake's bathymetry had become increasingly shallow due to 10 billion m^3 of sediment. During my field visit to the ULRP in September 2019, Agh explained that evaporation caused salt to stack up, layer up on layer on the bottom of the lake. As a result, a large area of the lake, especially the southern part, was 'prone to immediate evaporation' (Agh, 2014: 178). Accordingly, the lake area should be reduced to 3,800 km^2 while '100 km^2 [would be] reserved for *Artemia* and birds' (Agh, 2014: 178). In addition, salt and minerals should be extracted from Lake Urmia's northern part. The main goal of Agh's endeavour was to increase Lake Urmia's water level *and* its depth in order to reduce its vulnerability to evaporation and thus the unnecessary loss of water, as the tributaries' water would spread over a large, flat salt playa and evaporate quickly. To understand how Agh came up with his rather technical and highly anthropogenic intervention, it is important to stress his premise. From his point of view, the Lake Urmia crisis was a simple *matter of calculation*: how much water enters the lake, and how much leaves it? He assessed that evaporation would decrease the lake's water level by one metre annually, without considering water input. As long as the water depth remained—by reducing the area—above one metre, the lake's volume would increase annually. Only one of the analysed papers (see Barhagh et al., 2021: 27) supported this solution. In contrast to Kardovani, Agh proposed saving Lake Urmia albeit in a new, *manageable* form. Lake Urmia would not be fully restored to its pre-crisis state, but in this sense, he admitted that it would be best to accept this mutated version of the lake and adapt to its new status quo.

A third group was convinced that primarily water transfer projects from adjacent basins and the Caspian Sea would cope with the increasing water budget deficit. This perspective has already been discussed above as much as the fourth group. Their proposal included demand-focused solutions such as reducing irrigation water, stopping dam construction and prohibiting groundwater and surface water exploitation (Salimi et al., 2019: 615).

The described solutions differ significantly and sit at different ends of the spectrum in terms of ethics. Kardovani saw Lake Urmia as a dying lake, whose new form—salt and salt winds, more specifically—was a threat to the hydro-social collective and as such declared it an enemy. Instead of restoring the lake, it should be abandoned and combated by substituting it with a greened national park. At the same time, Kardovani strictly disagreed with proponents of both supply- and demand-focused solutions, as they would be politically and technically challenging, damage the agricultural sector and thus threaten people's livelihoods. Agh, similarly to the supply-focused solutions, relied heavily on the success of grand hydrologic infrastructure—in this case, a wall around a smaller area of the lake. However, he did not seem to engage in the discussion on whether demand- or supply-focused solutions would be more favourable to solve the crisis. Agh's proposition focused on morphing Lake Urmia in order to make it more manageable, and without this technical intervention, any measure to save it would fail. Therefore, he demanded acceptance of Lake Urmia's regression.

The demand and supply-focused solutions contrast these two above-mentioned solutions, in that they are convinced that the lake can be fully restored or revitalised. The main difference between these two is that supply-focused solutions wish to maintain or even increase agricultural production *and* provide Lake Urmia with enough water, whereas demand-focused solutions wish to restrict water use, especially in the agricultural sector, which, in order to survive, must adapt to the hydrologic conditions and surrender water for the sake of Lake Urmia's survival.

Against this backdrop, Kardovani's proposal was diametrically opposed to the other three groups. When faced with the question: to save or not to save Lake Urmia, he answered with the latter. However, neither 'letting go' of Lake Urmia nor Agh's techno-hydrologic intervention was accepted by the other experts. In the end, both demand- and supply-focused solutions prevailed and helped form the ULRP in 2013 under the lead of the Urmia Lake Restoration National Committee, chaired by the Vice President Jahangiri and the head of the Department of Environment, Isa Kalantari. Originally, the MoE was supposed to be in charge, but stakeholders and experts complained and accused the MoE of being responsible for the Urmia Lake disaster due to its aggressive hydraulic policies and dam construction (Salimi et al., 2019: 615).

7.5 ULRP—Its Goal(s) and Technocratic Re-organisation of the Basin

The ULRP (Urmia Lake Restoration Program) has already been criticised by Salimi et al. (2019) in terms of integrated water resources management, its organisational structure and performance. Therefore, I will focus specifically on the goals and solutions proposed by the ULRP, as some of these measures have generated heated debates in both the public and he scientific community about the effectiveness of this institution. I will also examine some of the management principles and objectives of the ULRP and compare them with those of its predecessor, the MoU. Four aspects will be elaborated: the ULRP's overarching goal, its water management paradigm, some of its most controversial policies and, finally, an overview of why the ULRP has failed to achieve its goals.

From the outset, there are significant contrasts between the ULRP and the MoU's plans to restore Lake Urmia. The MoU, as mentioned above, defined a 25-year vision and a goal, which aimed at eco-based water management in co-operation with all stakeholders. The ULRP has, or rather had, only one goal: to reach the ecological water level of 1,274.1 m by 2023, which it would achieve in three steps. First, from 2014 to 2016, the water level would be stabilised—simpler, stop its downward trend. Second, from 2017 to 2022, implemented projects would increase the water level. Finally, and third, when the ecological water level was reached, measures would ensure that this level was stabilised (ULRP 2018: 74). Whereas Sima et al. (2021) pointed out that the ULRP lacked multiple indicators to measure its success, I wish to emphasise how it approached its goal and could not achieve its overarching objective.

Issues with the Minimal Ecological Water Level The advantage of having only one measurable indicator is that everything falls in line to meet nothing but the designated goal, and hopefully this ensures Lake Urmia's revival—this is the means of the water level anyway. The issue, however, is that it misses everything else, which is not subsumed in this goal (Sima et al., 2021). More to the point, what if even the goal in itself could be corrupted and it failed to be a means to an end? In a Latourian (2018: 529) sense, the restoration goal, the water level, is a fiction: it is an actor, at a certain time (2023), in a certain place (Lake Urmia), that promises to revive a now dying lake. Everything was organised in accordance with this goal. Certainly, the water level would have to be—and was constantly—measured to record the progress of the programme, and this demanded a reliable scientific 'chain of reference' to Lake Urmia—everything stood and fell, so to speak, with a curve that potentially rose and fell like a stock market index, hopefully never to hit rock bottom.

However, the ecological water level has become a dispositive, a target whose original intention has been lost or has remained at least unmentioned in the ULRP (2015, 2018) reports. With this in mind, with which entities did the water level hope to establish a connection in the first place?

The purpose of the ecological water level was to indicate a sustainable reproduction rate for the local brine shrimp *A. urmiana*, because a higher water level suggested a tolerable salt concentration for this species. In other words, it was hoped that a larger population of *A. urmiana* would solve many issues, as already discussed in Chapter 5.1. The issue is that the minimal (!) ecological water level turned out to be an indicator for an indicator for an indicator; one might say different grades of indicators, namely those closest to the brine shrimp and those furthest away (Fig. 7.7). The water level (fourth-grade indicator) should indicate a certain volume of water (third-grade indicator) that sets up salt concentration (second-grade indicator) for a sustainable reproduction of *A. urmiana*, expressed in cysts per litre (first-grade indicator). The latter three indicators would have to 'fall in line' with the water level in order to indicate a positive a progression—and ultimately Lake Urmia's revival. What is always at stake with the minimum ecological water level is that it can lose its *meaning*, if the lateral chain of references fails to be a conduit for *A. urmiana*'s sustainable reproduction. The water level must indicate a certain water volume, and this in turn must guarantee that salt concentration stays within the eco-friendly boundary; the tolerable salt concentration must enable *A. urmiana* to reproduce sustainably. Conversely, if any of these factors *change*, then successful referencing or indications from a concrete water level to Lake Urmia's 'health' are lost. Chapter 5.1 has already illustrated how even the closest indicator for the reproduction rate of *A. urmiana*, namely salt concentration, is a controversial topic. If other factors turn out to be variable, more volatile than presumed, then the water level would be nothing other than a meaningless number, something that once indicated Lake Urmia's health for a specific time, and a specific Lake Urmia, which has now lost its ability to do so. And indeed, so it has happened, but not due to the discussion on the water volume that would be necessary to reach the ecological water level, but for another reason. Scientists and even the ULRP (2018: 61) itself have recognised that the lake's bathymetry has changed over the years, meaning that, as indicated by Agh, it has become shallower due to sedimentation during this mutation. In this context, Saemian et al. (2020: 45) note:

> It is important to mention that the bathymetry of the lake is not constant during the study period, mainly due to the sediment of salt to the bed of the lake. Therefore, for a better evaluation, a bathymetric map of the lake for some other period of 2002–2019 would be needed.

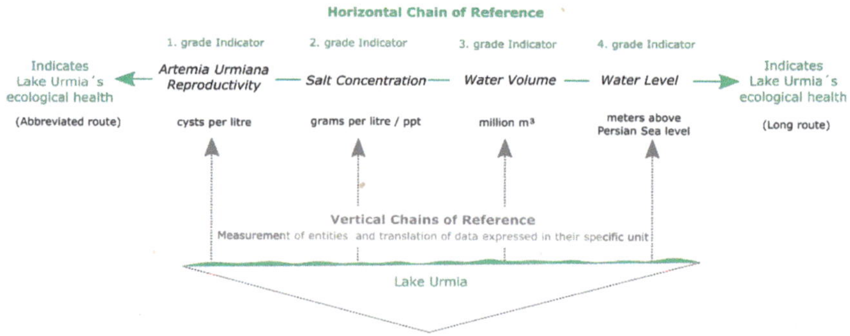

Fig. 7.7 Horizontal and vertical Chain of References for ecological water level

Hemmati et al. (2021: 2) furthermore complain that 'a detailed bathymetry of the lake bed was also not available to use in the modeling'. Consequently, Danesh-Yazdi and Ataie-Ashtiani (2019: 649) demand:

> This clearly underscores the necessity for a hydro-dynamic model so that continuous changes of the lake bathymetry in response to inflow and in-lake processes can be tracked. Regarding the complex lake bed bathymetry, validation of such a model requires intensive data collection describing the spatial distribution of salinity and flow velocity within the lake.

To put the situation bluntly: the thesis by Nazaridoust on the ecological water level tried to make an impossible calculation to define an optimum water level for Lake Urmia. However, this does not imply that the task was meaningless. On the contrary, it was driven by meaning, a means to an end, the revival of Lake Urmia. The issue is just that the ULRP, the MoU, the law and even the majority of scientists have neglected the ever-changing coefficients of this calculation. They have nothing but 'double-clicked' (Latour 2013: 93f), they have not scrutinised, transformed or optimised the impossible calculation, and as such the environment water level has become a dispositive, a fact and a hegemonic measurement for success. The ULRP has lost its scruples in relation to the minimum ecological water becoming immoral, which 'come[s] from the suspension of the reprise, from the abandonment of cases, indifference toward any technical arrangement of proofs. We stop worrying; we suspend our scruples. We use the rich infrastructure of organization to proclaim: "We're quits forever"' (Latour, 2013: 461). The absence of these scruples has been criticised by many authors (see Parsinejad et al., 2022: 9), who demand a spectrum for the water level, where the optimum possibly lies but remains incalculable in terms of an exact figure—or in other words, a spectrum provides scruples with more breathing space to settle uncertainty.

Reiterating the issue with the minimum ecological water level once more: identifying the optimum conditions for the *A. urmiana* and transforming this exact figure into a concrete water level is destined to failure, because the lake's bathymetry is dynamic, just as much as almost any other factor influencing the calculation (e.g., anthropogenic climate change, droughts, river inflow). The goal of the ULRP, to revitalise Lake Urmia, in concordance with a specific number might be, according to my argument, immoral, since the minimum ecological water level may never fulfil its promise of reaching the 'envisioned'" Lake Urmia and will thus fail to adjust its target. This is the immorality of this fixed number.

Accordingly, if the very goal of the ULRP itself is questionable and relies only on its organisational structure and the minimum ecological water level as a target that always faces the danger of being stripped of its meaning, then every measure that follows is touched by this vagueness. However, this would be unproblematic overall if the ULRP had defined a vision, a goal that was more abstract but more inclusive of other entities. In comparison, the MoU (2010) aimed at the minimum ecological water level as a sub-goal, not a main goal. Its target was ecology-based, sustainable water management, which included environmental protection and co-operation with local stakeholders. As we shall see, the ULRP departs from the MoU by some distance.

The 40% Reduction Controversy To reach the target water level by 2023, the ULRP devised six strategies (ULRP 2018: 81):

1. "Controlling and decreasing agricultural water consumption."
2. "Supplying water for the lake from new resources."
3. "Controlling and decreasing water withdrawal from surface and groundwater resources."
4. "Studies and software measures."
5. "Conservative and mitigation measures."
6. "Facilitating and increasing inflow to the lake through construction."

Four of these six strategies were concerned with redistributing and reallocating water either within the basin (Strategy Nos. 1, 3 and 6) or from new sources (Strategy No. 2). The other two strategies mainly targeted increasing the ULRP's understanding of Lake Urmia, determining and evaluating possible threats (e.g., causeway, salt winds) and other 'soft' solutions that will be elucidated later on. By the look of it, one might be tempted to indeed characterise the ULRP's strategy as sustainable, since it included the three pillars of sustainability: environmental, economic and social. The ULRP's budget plan for the fiscal year 2015/16, however, clearly emphasised the reorganisation or transformation of the agricultural sector and (macro) engineering solutions (Fig. 7.8). In total, 95% of the budget was planned to be spent on these two measures.

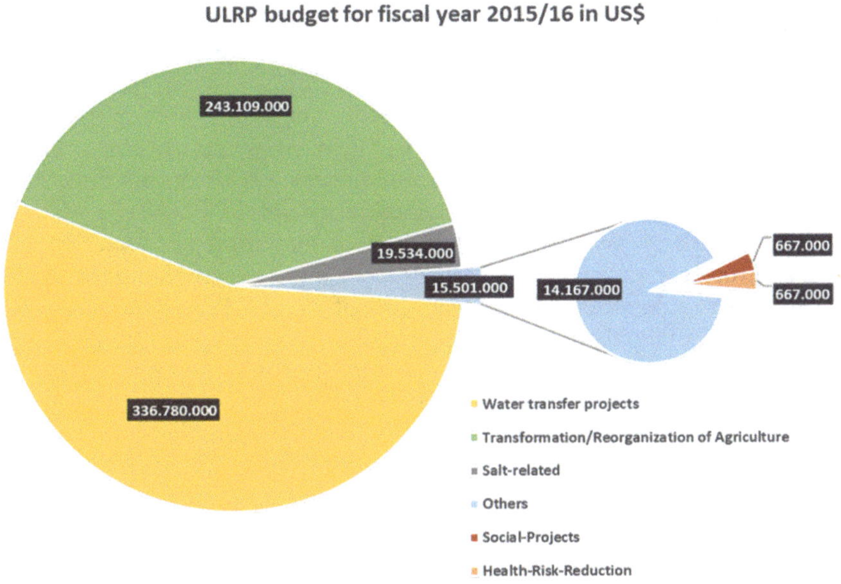

Fig. 7.8 ULRP budget plan for the fiscal year 2015/16 2015 (*Source* ULRP)

The only social projects included in the budget were the study of alternative livelihoods and employment for farmers (US$ 667,000), along with the reduction of health risks associated with the lake's desiccation (US$ 667,000). The ULRP even planned to invest more into public education and awareness campaigns (in total US$ 6,667,000) than in social projects. This is not to say that the ULRP should have invested an equal amount into social projects as it did for water transfer projects or in the agricultural sector but that it underestimated the social nature of its entire endeavour. Especially, the control and regulation of agricultural water was a severe intervention in farmers' livelihoods that was often unwelcomed. As such, the ULRP should have paid more attention, and therefore more money, to social, or more specifically farmer-related, projects. There is—as mentioned several times—an insufficient number of academic research on the social dynamics at Lake Urmia. Despite the ULRP's positive, catalysing effect of intensifying academic interest in the matter of concern, it recognised that it lacked the adequate capacities to conduct social studies (Tussupova et al., 2020: 16). The ULRP may have perceived its intervention as *water resources management*, but its alterations concerned the hydro-*social* collective at Lake Urmia—including farmers *and* their farmland.

Although social aspects received little attention, they were taken into account in the ULRP's plan. In their 2015 report, for example, they assumed—by default—that by implementing their programme 'local dwellers [would] be the main beneficiaries' (ULRP, 2015: 18). In contrast to the MoU (2010), however, co-operation between authorities and locals, e.g., farmers, was marked by a top-down hierarchy, in that locals were understood as an extension of the ULRP actions, and the people were obliged to implement the plans to save Lake Urmia. Moreover, each and every individual was called upon to 'monitor and report illegal practices' by others (ULRP, 2015: 19). In order to restore the lake, everyone—starting from the bureaus—must fall in line, even if it meant potentially disrupting social cohesion in rural areas. The ULRP wanted farmers to be the extended arm, the very fingers, ears and eyes of their plan, while the institution remained the head and brains of the operation. This top-down scheme was not only relevant in the context of excessive groundwater extraction, but in the overall plan of the ULRP to emphasise drastic demand-focused solutions.

One of the ULRP's most controversial measures was to reduce agricultural irrigation or consumption, by 40% over five years, or 8% per year, to save 25% of renewable water resources (ULRP, 2015: 23)—a plan that included 'boosting' irrigation efficiency (ULRP, 2018: 82). According to my interviews with farmers and experts, the ULRP's goal was met with a mixed reception.

During my field visit in September 2019 in the adjacent villages around Lake Urmia, only six out of 70 farmers used drip irrigation. One farmer even stated that his motivation to use modern technology was directly related to Lake Urmia's crisis:

> I am afraid of the lack of water [...] I want to save Lake Urmia'.

Those who applied drip irrigation reported positively in terms of increased crop production. However, as most farmers did not use modern irrigation methods, I asked them why they had rejected them. They gave me three main reasons: lack of money, lack of support from the government and occasionally mistrust in modern technologies. Most farmers in the Lake Urmia basin of West and East Azerbaijan are small-scale farmers, whose cultivated land covers an average an area of only one to five hectares (ULRP, 2018: 56). They expressed that due to their small plots of land, they had no interest in changing their irrigation techniques and even stood behind their traditional surface irrigation. In their experience, the new irrigation methods were 'not good for the land' or, more precisely, it was only feasible for vegetables and not for trees. In one village, a farmer's neighbours changed to drip irrigation, but it did not work for them, as their trees apparently could not adapt to the new method. Thereafter, they reverted to their former, traditional irrigation approach.

The majority of interviewed farmers could not afford modern irrigation technologies and hoped for government support, which was absent in almost all cases.[3] Interviews conducted in 2020 revealed that most farmers may have possibly welcomed new irrigation methods, but they simultaneously needed economic support, if they were to save water as the ULRP intended to achieve with its measures.

> If the government covers the running costs of the farmers to some degree, they could save some of the water resources. (Farmer-1)

Next to this, other farmers added that reducing irrigation use by 40% was possible:

> But under one condition: state institutions should constantly control everything. The economic losses that may result for farmers and ranchers should be taken into account and compensated. In this case, it is even possible to save much more water. (Farmer 2)

> The government should support farmers and teach them modern methods of cultivation so that they can achieve better results than conventional [traditional] cultivation. (Farmer-3)

These insights confirm Abbaszadeh et al. (2020) and Farahmand et al.'s (2023) findings that farmers in general lay responsibility for the issue at the government's door. In general, farmers would welcome new

[3] Only in one village did the government subsidise 85% of the costs, while the village paid the other 15%.

irrigation methods, but they complained that these were too expensive without any financial support.

Conversely, some farmers expressed their mistrust in the ULRP project:

> Unless those responsible for the farmers can *justify* what key claim is behind this figure [(40% reduction of irrigation water)], this figure does not seem reasonable. The [financial] support somehow disappears and we do not have that much of it. In the end, we feel the problem of water and drought much more than the authorities. (Farmer-3)

Indeed, the ULRP (2015: 11) did not disclose how this figure of 40% had been calculated but referred to the Ministry of Agriculture and the Ministry of Energy as the authorities 'enforcing' this measure. The project included that next to the implementation of new irrigation methods, the Ministry of Energy would purchase the farmers' water rights within five years (ULRP, 2015: 23; ULRP, 2018). However, none of my interviewees was contacted by any authorities during my time in the field from September to October 2019—neither for this nor on any other matter. Without financial support or a government-run purchasing system, the endeavour would have likely caused social distress (Danesh-Yazdi et al., 2019: 650).

In the scientific community, the ULRP's demand-focused measure aimed at the agricultural sector was similarly perceived and evaluated heterogeneously. AghaKouchak et al. (2015: 311) emphasised that such policies were necessary despite losses in the agricultural sector. In their own words, 'they are essential to sustain environmental and economic productivity of the Lake Urmia basin in the long run' AghaKouchak et al. (2015: 311). Hamidi-Razi et al. (2018: 10), based on their 'numerical modelling approach' and a two-dimensional depth-averaged model of Lake Urmia, argued that the 40% reduction in agricultural water consumption would indeed significantly improve and reach the minimum ecological water level in four years. Similarly, Zarghami and AmirRahmani (2017: 322f.) concluded that increasing irrigation efficiency would contribute the most to Lake Urmia's restoration. Barhagh et al. (2021: 25), for their part, quantified that improving irrigation efficiency by only 21% would increase Lake Urmia's water level by ca 2 metres within 10 years.

Other research, however, took a more critical stance on this restoration measure. Alborzi et al. (2018: 8f.) proposed a 20% reduction in water consumption as a more realistic scenario, which would, depending on climatic conditions, take three to 12 years for Lake Urmia to reach its ecological threshold. The ULRP would be 'overambitious' with its timeline if it tried to implement its 40%-reduction programme (Alborzi et al., 2018: 7). Moreover, Alborzi et al. (2018: 9) suggested using the so-called 'environmental hedging' approach developed by Adams et al. (2017: 7830): 'Environmental hedging 'hedges bets' for drought by reducing releases for fish, sometimes intentionally killing some fish early to reduce the likelihood of large fish kills and storage crises later'. Although Alborzi et al. (2018: 9) did not mention how this approach could be applied at Lake Urmia, they insisted on a partnership ethics solution to 'facilitate progress towards ecological recovery of the lake while meeting human demands within the constraints of basin scale water availability and ecological functions of Lake Urmia'.

Shadkam et al. (2020: 14) drastically contradicted the irrigation efficiency paradigm. According to their study, the ULRP's goal to reduce water demand in the agricultural sector by 40% would decrease water withdrawal but increase depletion. In other words, instead of saving water, the measures would increase agricultural water use. There are manifold reasons to explain this rebound effect. According to Shadkam et al. (2020: 11;14), this was due to future population growth, socio-economic development and less water availability due to anthropogenic climate change. Moreover, the agricultural sector was already being impacted by water shortages—or anthropogenic shortages—which meant that water demand could not be met by current water withdrawal. Therefore, even if more efficient irrigation technology did actually reduce water demand by 40% and water withdrawal by 23%, water depletion—water that 'lost' from the basin—would still increase by ca 17% (Shadkam et al., 2020: 14). This was mainly due to a decrease in surface and groundwater return flow, as drip irrigation, for example, only focused on the crops' water needs, and so less water would be applied to the soil (e.g., flood irrigation) to recharge groundwater and return to surface water.

The issue with modern irrigation such as drip irrigation has been explored in detail by Perry et al. (2017) (see also Grafton et al., 2018), who published an FAO report on the topic. According to their report, more efficient irrigation technology does not fulfil the usual win–win-expectation; that is to say, to conserve water and increase crop production per unit of water. Based on 150 addressed experts, Perry et al. (2017: xi) summarised:

> […] [T]here are rather few examples of carefully documented impacts of hi-tech irrigation, while there are many examples of projects and programmes that assume that water will be saved and productivity increased. Such studies as do exist, are either inconclusive or, more often, show that water consumption actually increased (as science would predict) when irrigation systems were upgraded, and that productivity per unit of water consumed was more or less constant.

In contrast to the ULRP, Shadkam et al. (2020: 14) emphasised and disclosed a different understanding of irrigation efficiency. The ULRP relied on the concept of *classical efficiency* for their strategy, which indicates how much of withdrawn water has been used for crop production. This in turn reduces the indicator to a simple in and output relation (ideally, water withdrawal equals water demand, which would be a 100% classical efficiency). The suitable indicator to evaluate the restoration progress for Lake Urmia would have been the *effective efficiency* according to Shadkam et al. (2020) because return flow would also have been considered. These two concepts changed the perspective on efficient irrigation significantly. According to the ULRP (2018: 80), the decrease in water demand should be accompanied by a 'boost' in irrigation efficiency from 40 to 75%. Classical efficiency implies here that agricultural water use is 'insufficient' from an economic standpoint, because water is 'wasted' instead of effectively used for crop production (e.g., due to flood irrigation). This concept, however, fails to account for real water savings, in the sense of how much of the irrigated water *returns* to the basin's water table. Instead of comparing total water withdrawal to the water used for crop production, effective efficiency considers how much water was effectively used for crop production minus the return flows. As such, Shadkam et al. (2020: 14) assumed a 75% effective irrigation efficiency in the Lake Urmia basin, which was relatively high and did not leave much space for improvement. This perspective contradicted, next to the ULRP, Madani's (2014: 320) criticism of Iran's poor irrigation efficiency (35%, while only 5% of irrigation was based on pressurised irrigation), demanding that it should 'modernize the agricultural sector' (see also Madani et al., 2016: 1000; Barhagh et al., 2021: 23).

Shadkam et al. (2020: 12) calculated the overall low effectiveness of the agricultural strategies planned by the ULRP. Other measures, such as inter-basin or wastewater transfers, were more effective in reaching the desired ecological water minimum (however, Shadkam et al., 2020 considered the suggested water volume of 3,100 million m^3 calculated by Abbaspour and Nazaridoust (2007) as the minimum threshold). In the face of the large investments and high prioritisation of the ULRP on reorganising the agricultural sector, this analysis seems detrimental. Against this backdrop, Shadkam (2017: 97) expressed in her dissertation that:

> [i]n general, water supply-oriented measures such as inter-basin water transfers are seen as the least supported measures. This is surprising, as the quantitative assessment showed that water supply–oriented measures would not only increase inflow to the lake the most, but that the success of ULRP also depends on them. [...] [W]ater availability of the basin will likely decrease due to climate change. Therefore, water supply–oriented measures should be taken into account [...].

Therefore, the second controversially discussed solution, water transfer projects, will now be elucidated.

7.6 Changing the Flow of Rivers

In February 2023, Iran's new President Raeisi inaugurated a 'mega-project' to transfer water from the Kanisib Dam reservoir (little Zab River) to Lake Urmia. During his inauguration, Raeisi clarified that he was a strong proponent of supply-focused solutions: 'The restoration of Lake Urmia is a national concern, and the water transfer project is the most important plan in this regard' (Presstv, 2023). He furthermore politicised the project by stating that '[t]he enemy does not want us to gather together here in complete security, with unity and cohesion and celebrate the inauguration of a national environmental plan' (Mehr News Agency, 2023). In total, 36 km of tunnels and 11 km of canals were planned to transfer 600 million m^3 water annually to Gadar River in order to restore the lake's minimum ecological water level by 2028 (*Tehran* Times, 2023). Originally, the water transfer project was supposed to start in 2019, but according to Salimi et al. (2019: 621), the ULRP allocated only 30% of its planned budget from 2014 to 2017. Nevertheless, the responsible authorities continued to rely on expensive water transfer projects in order to reach their designated goal. Next to the Kanisib Dam, additional water would be transferred from the Silveh Dam water reservoir, constructed on the Lavin River, which is also located in the transboundary little Zab basin between Iran and Iraq. A 21-kilometre-long channel was planned to provide 190 million m^3 to Lake Urmia (ULRP, 2018: 86).

For the ULRP's restoration plan, these inter-basin water transfer projects were second only to the reduction in agricultural water use, with a total supply of 690 million m^3 to 1,340 million m^3, respectively, in 2023. Intra-basin water transfer projects, which included the dredging of rivers, opening of dams and transferring recycled wastewater from urban areas, particularly Tabriz, would provide 550 million m^3 in 2023, according to the ULRP's plan.

At least two binary points of opposition can be identified in the controversy over water transfer projects at Lake Urmia and in Iran in general, especially inter-basin transfers. On the one hand, there are its proponents such as the central government, the Ministry of Energy, the ULRP and a few scientists. With a few exceptions, such as Golabian (2011), these proponents include demand-focused solutions next to water transfer projects, but they see the latter as a necessity to address the urgency of Lake Urmia's demise and to adapt to anthropogenic climate change, as well as future socio-economic development. However, not every water transfer project is perceived or assessed as equally reasonable. Therefore, there are different groups *within* the proponents of water transfer projects. On the other side of the spectrum lies the opposition, who thinks that water transfer projects cause more harm than good. Environmentalists, the Department of Environment and the majority of scientists contradict such supply-focused measures. Yet again, though, some scientists are prepared to tolerate water transfer projects under certain conditions. To get a better grip on this controversy, I will now focus on the arguments offered by the opposition, as this perspective has not been elucidated so far.

The arguments can be categorised as environmental, economic, political, psychological, technical, ethical and ideological. Some scientists oppose water transfer projects for ideological reasons because they resemble, and therefore continue, the root cause of the Lake Urmia crisis: the hydraulic mission. Madani et al. (2016: 1005) insist that Iran's hydraulic mission is still ongoing; that is to say, leading authorities persist with ever-new constructions or technologies that control water rather than manage it (Madani, 2014: 321). Although dam constructions have come to a halt—especially in the face of desiccating water bodies and political tensions—other projects such as water transfer and desalination plants still bear the imprint of the hydraulic mission:

> The water mafia is making a shift in their policies and in their interests; they are turning their attention to water transfer projects. Now, they are working on water transfer projects as a substitute for their hydraulic mission to make profits again and to get the money for those projects. (expert-5)

One could say that the moral implication of the hydraulic mission and its technological infrastructure, which is entirely anthropocentric, is impregnated by the authorities' desire to produce the most economic value through its water resources. In simpler terms, applied technologies carry a moral weight that is in synchrony with a farmer's desire to increase his or her economic income. Water, accordingly, is abstracted to its economic value, in that it flows towards money, as the scientists of the hydro-social cycle would argue. Particularly inter-basin water transfer projects would perpetuate the hydraulic mission's immoral implications that antagonise the environment, as is the case in the Lake Urmia basin (Hashemi, 2012: 164). Technologies propagating this immorality to water users would be politically inconclusive with an entity such as Lake Urmia—in the Politics of Nature, they could not live together, to put it in Latourian terms. Therefore, the 'quick fix' of water transfers (AghaKouchak, 2015: 310) would quickly turn into a short-sighted 'solution' that would continue the hydraulic mission instead of changing the water management paradigm to a sustainable one (Madani, 2014: 319). The MoU (2010: 59), for example, saw a conflict or even contradiction in reconciling sustainable, ecocentric management with inter-basin water transfer, as such measures stem from or co-produce a morality that is diametrically opposed to the preservation of a healthy environment. Water transfer projects suggest abundancy for water users in the donor basin and actually lead to more water consumption—the argument in a nutshell (Madani et al., 2016: 1012).

In this context, scientists, environmentalists and the Department of Environment warn of the environmental consequences of implementing water transfer projects. Here, two distinct water concepts can be identified. While Golabian (2011) emphasises the 'relatedness' of Lake Urmia and her 'elder sister' the Caspian Sea—which resonates the idea of *universal nature*—albeit opponents of these mega-projects highlight the differences between the two distinct water bodies as an insurmountable obstacle. In short, modern water—homogenous, ubiquitous, transferable—is contrasted by particular *waters*—heterogeneous, specific, local.

However, I must stress, that the 'clash' between these two water ontologies is disproportional in the scientific community. Except for Golabian (2011), none of the other analysed scientists expressed their water ontology so openly. Therefore, if I were to immediately ascribe the modern water ontology to every scientist who advocates water transfer projects, it would artificially polarise the discourse, which I do not intend to do. After all, what most scientists think about water can rarely be concluded so clearly from the papers that I could push them to one or the other extreme. However, those who propagate water transfer projects without hesitance could internalise modern water. The reason being is that several experts warn that the particularities of different water bodies, such as Lake Urmia and the Caspian Sea, should not be mixed without assessing the possible negative side-effects in advance. The botanist and environmentalist Hossein Akhani, for example, uses an anthropomorphic example to make this point. In his comparison, the transfer of water from other basins to restore Lake Urmia is similar to a blood transfusion. He warns:

> [...] there are factors determining the right type of blood for each person, likewise we cannot transfer any water from anywhere to quench the Lake Urmia. [...] Each basin has its distinguished physical and biological characteristics and inter-basin water transfer can have unpredictable effects on the lake ecosystem such endangering the biodiversity of the region [...]. (*Tehran* Times, 2018)

In this context, the Caspian Sea especially has been cited by several stakeholders as being potentially harmful to Lake Urmia's ecology. The former East Azerbaijani MP Ali Alilu from the Conservative Party, characterised the Caspian Sea as polluted, and hence dangerous to the lake because oil and heavy metals would be transferred into it (Salimi et al., 2019: 631). Khatami and Berndtsson (2013: 6) from the Department of Water Resources Engineering in Lund University, Sweden, also stated that using water from the Caspian Sea would 'affect the whole ecosystem through its path to Urmia Lake. [...] [Thus], common sense would suggest that for solving a problem one should not introduce new problems'. While the UNEP & GEAS (2012) agree that the Caspian Sea would be indeed problematic, they suggest transferring water from neighbouring basins. However, the Aras seems to be similarly polluted by industrial and mining activities in the upper riparian countries of Armenia and Azerbaijan (Ebrahimi et al., 2023). Given these arguments, the former head of the Department of Environment, Masoumeh Ebtekar, declared the water transfer from the Caspian Sea or Aras River as 'unprofessional' (Rajabova, 2013). In order to avoid possible environmental damage, AghaKouchak et al. (2015: 310) called for a detailed analysis that would determine the characteristics of the two different waters—and therefore their compatibility. The transfer of foreign substances in the water and the biogeochemical impact must always be considered in water transfer projects. De facto, the plurality of water is emphasised here, which is clearly distinguished from a possible reduction as mere, homogenous water. In this sense, the attitude of the opponents of such projects is always a political one, in that they reject the view on water that propagates that a transfer from one water body to another can be carried out without any concerns.

At the same time, these supply-focused strategies are often criticised for their high costs as well as their potential to cause unintended political tensions with neighbouring countries. The Caspian Sea, for example, is surrounded by five littoral states (Iran, Azerbaijan, Russia, Kazakhstan and Turkmenistan), and the Aras is a transboundary river shared between Turkey, Armenia and Azerbaijan (Madani, 2014: 318). Using these international water bodies to restore Lake Urmia would not only be politically sensitive, but it would furthermore be accompanied by the concern regarding being dependent on the goodwill of neighbouring states—and thus risk being more vulnerable, so a member from the Ministry of Energy indicated (*Tehran* Times, 2018). Implicitly, this means that if the Urmia basin were to expand its water sources significantly, the political integrity of the region would always be at risk of collapse.

From an effectiveness perspective, most studies note a marginal and an even negligible role of water transfer projects—this goes for inter and intra-basin projects. Zarghami and AmirRahmani (2017: 324), for example, calculated that water transfer from the Zab and Aras rivers would contribute only 4% in terms of reaching the minimum ecological water level. Similarly, Hamidi-Razi et al. (2018: 8) saw 'no significant contribution' made by transferring water from the Zarrineh River to the Simineh River.

As a result, the ULRP's goal to rely heavily on water transfer projects contradicted the hegemonic perspective in the sciences at Lake Urmia. Figure 7.9 summarises the ULRP and its planned water transfer via different methods. Next to the intra-basin measures, for example dredging rivers, inter-basin projects also play a significant role, because if these volumes of water are not directed to the lake, the task is destined to fail. Furthermore, note in Fig. 7.9 that the ULRP's restoration plan assumed an *increase* in the natural flow of rivers, which beg the question, what was their understanding of 'natural' in terms of river flow? Would this inflow increase as a consequence of fewer illegal wells drilled in the basin, the successful implementation of demand-focused solutions? If not, then the increase in water volume in rivers seems questionable in the face of the increasing effects of anthropogenic climate change, as expressed by Shadkam et al. (2017).

Even more so, there arises another issue with these assumptions made by the ULRP. A commonly discussed issue of concern within the scientific community and other reports is the general lack of data, for instance with regard to the basin's water budget, which renders measures such as this one arbitrary and open to interpretation (see e.g., Parsinejad et al., 2022; Sima et al., 2021; Danesh-Yazdi & Ataie-Ashtiani, 2019; Chaudhari

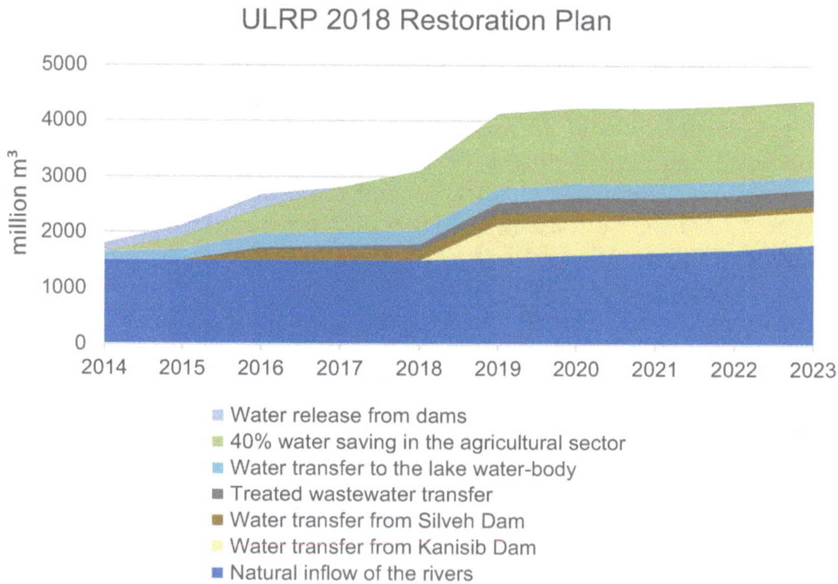

Fig. 7.9 ULRP's plan to restore Lake Urmia by 2023 in million m^3 2018 (*Source* ULRP)

et al., 2018; Vahedoost & Aksoy, 2018; Delju et al., 2013; Khatami & Berndtsson, 2013; Hashemi, 2012; Hassanzadeh et al., 2012). In a common report issued by the FAO & ULRP (2017: 17) it was argued that:

> The data collected in Lake Urmia does not include water consumption at the field level, and the limited data that exist are either out of date or deposited in inaccessible archives. Relying only on guesstimates for water consumption or availability is not good enough for Lake Urmia restoration purposes. This is a significant data gap.

In their 2019 study, Danesh-Yazdi and Ataie-Ashtiani employed a polemical approach, as evidenced by the title of their work, 'Lake Urmia Crisis and Restoration Plan: Planning without appropriate data and models is gambling'.

7.7 Conclusions: Lake Urmia and its Restoration at a Dead-End

From 2016 to 2019, Lake Urmia experienced an upward trend. It seemed that the restoration measures put forward by the ULRP were starting to bear fruit. Flamingos returned in greater numbers (10,000 in 2017 and 35,000 in 2018; *Tehran Times* 2017, 2018b) and the water level rose to what it was in 2011, which was before the lake almost disappeared in 2013. The media celebrated the restoration process, proclaiming its 'revival', 'rising water levels' and how it had been 'brought back from brink'. The statements went even as far to say that 98% of the lake had been restored and the minimum ecological water level would be reached by 2022 (*Tehran* Times, 2021). Even the UN-Environment foresight brief in 2017 reported on the 'signs of recovery', related to satellite images showing an increasing surface area compared to 2013. The worst—so it appeared—was left behind, the crisis had reached its deepest point and Lake Urmia, like a stock price, had escaped from its depression. Lake Urmia was finally stabilised—restoration and revival now at least possible. But not for long. Since 2022, the media has continuously published articles stating that the lake is 'on the verge of disappearing' or about to 'perish'. Indeed, the short-lived upspring of Lake Urmia might have been caused by a rain anomaly from 2016 to 2019 and less so by the restoration efforts of the ULRP (Nikraftar et al., 2021). Although I stressed at the beginning of this chapter that Lake Urmia and the Aral Sea should not be compared as if they were victims of the same processes, by all means they share the same fate: mutation.

The Lake Urmia that used to be is no more.

What has entered the political stage, intruded upon the fields of adjacent farming land via salt winds, caused depression or the disentanglement of people from their environment is a new, mutated lake—a lake that, according to the sciences, media and government, has the potential to threaten public lives and disrupt social cohesion in the region and possibly beyond. The dissonance between what Lake Urmia used to be and what it has become is in itself an actant. There is a clear discrepancy when we think of Lake Urmia, namely its former potential (e.g., Artemia industry, UNESCO Reserve) and the new salt desert surrounding the rest of the lake. It is the uncertainty about what the future holds for the Lake Urmia collective, one that experts and statistics wish to narrow down to probable outcomes. It is similar to what Madani said when he criticised the control paradigm in Iran's water management. To control water is delusional. It has a 'life' of its own, and it is all too capable of evading any merit to regulate it by some ideal(ised) design. Change or metamorphosis, the process of becoming is always flanked by past memories, present uncertainties and future projections. Lake Urmia of the 1980s to mid-1990s has disappeared in the face of the basin's transformation, while its new, threatening form has shaken the entire region and the central government.

In contrast to the common assumption, however, that Lake Urmia has been the victim of the Aral Sea Syndrome, I have argued that we should be aware of the potential fallacy of such a view. There are manifold processes at work in the Lake Urmia basin, numerous disruptive forces, whereas the Aral Sea mutated as the result of a clear, linear—at least economically—and successful agenda developed by the Soviet Union. Lake Urmia's desiccation is the result of failure. And as such, the lake has been 'torn apart' due to anthropogenic particularism and climate change.

The solutions to Lake Urmia's restoration are controversial (Fig. 7.10); however, this chapter has shown that in the sciences there is consensus among the majority of researchers that in order to 're'-morph Lake Urmia, at least to its minimal baseline, water consumption must be reduced. Therefore, demand-focused solutions are at the top of the agenda, especially for natural sciences and the Department of Environment. A minority of scientists, the central government and the Ministry of Energy insist that supply-focused solutions are mandatory to save Lake Urmia. The demand-focused parties wish to adapt and transform the agricultural sector primarily, whereas additional, alternative solutions are proposed that target increasing people's livelihoods and capacity to cope with economic and ecological issues. These alternative solutions are not directly aimed at altering the hydrological cycle but at the hydro-social cycle, since an increasing number of scientists stress that we must be aware of the social barriers to Lake Urmia's restoration.

Moreover, the comparison between the Memorandum of Understanding (MoU) and the ULRP's roadmap disclosed diverging ethical stances as well as different priorities. While the MoU emphasised a long-term goal that envisioned an aesthetic, populated and healthy socio-environment, the ULRP narrowed down its main target to reach the minimal ecological water level. Both roadmaps were characterised by an ecocentric ethical position. However, the difference between the two lies in the fact that the MoU slightly leant towards a partnership ethic, since it involved local stakeholders and their needs, whereas the ULRP focused almost entirely on the lake's health and neglected any anthropogenic circumstances. The ULRP's goal to reach the ecological water minimum—originally—by 2023 failed perhaps mainly due to a lack of financial commitment by the central government and—as I argued—due to their narrow-minded target. Indeed, the ULRP's homepage was shut down in December 2023, and no more information is available on the future of the organisation.

The water concepts that were part of the restoration discourse revealed that modern water does prevail in practice. Most solutions presented by the ULRP were technical in nature. Both new irrigation technologies and the substitution of local/traditional knowledge with modern, scientific know-how mark a distinct characteristic of modern water, since they are usually perceived as helpful by default. Although environmental organisations or experts disclosed an additional water concept that is more heterogeneous and does not favour simple, technical solutions such as water transfers, especially from different basins, the hegemonic water paradigm still propagates modernisation as a hammer for all nails.

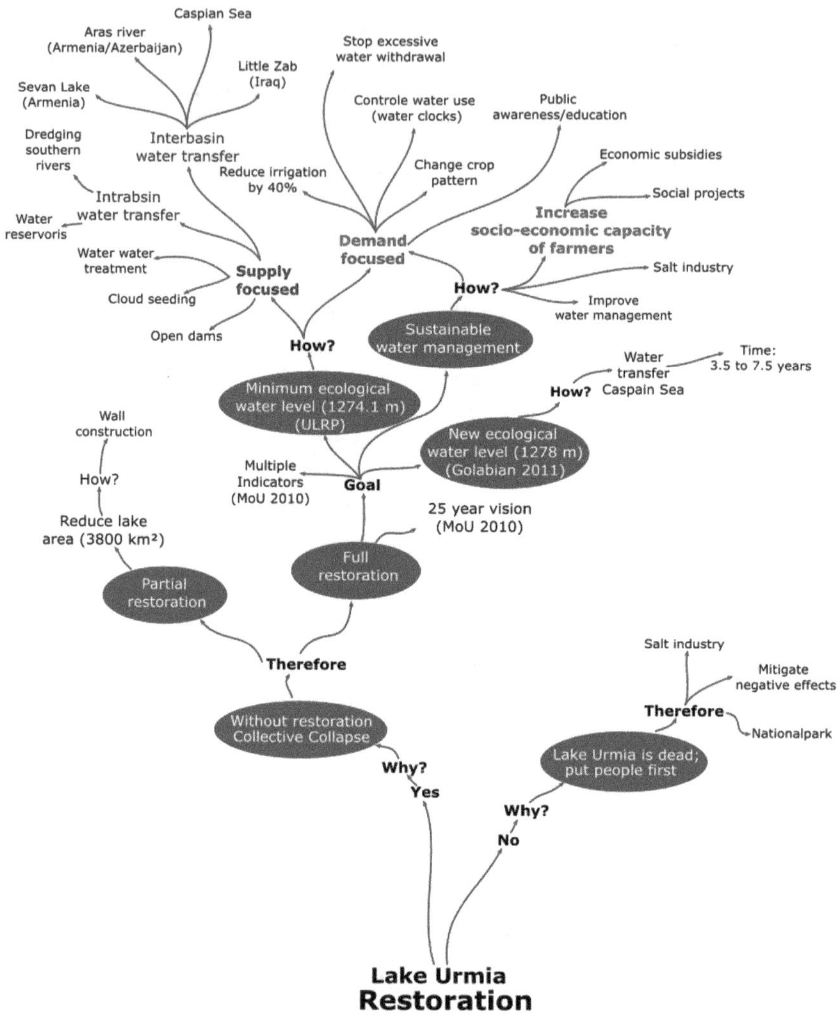

Fig. 7.10 Tree of Disagreement on Saving Lake Urmia

References

Abbaspour, M., Javid, A. H., Mirbagheri, S. A., Ahmadi Givi, F., & Moghimi, P. (2012). Investigation of lake drying attributed to climate change. *International Journal of Environmental Science and Technology, 9*(2), 257–266.

Abbaspour, M., & Nazaridoust, A. (2007). Determination of environmental water requirements of Lake Urmia, Iran: An ecological approach. *International Journal of Environmental Studies, 64*(2), 161–169.

Abbaszadeh, M., Banifatemeh, H. & Nikdel, N. (2020). A Study of Farmer's Perception of Urmia Lake (UL) Crisis with Grounded Theory Approach. *Journal of Applied Sociology 31*(2), 1–4.

Adams, L. E., Lund, J. R., Moyle, P. B., Quiñones, R. M., Herman, J. D., & O'Rear, T. A. (2017). Environmental hedging: A theory and method for reconciling reservoir operations for downstream ecology and water supply. *Water Resources Research, 53*(9), 7816–7831.

Agh, N. (2014). How to save the dying Lake Urmia. *Acta Geologica Sinica, 88*(1), 178–179.

AghaKouchak, A., Norouzi, H., Madani, K., et al. (2015). Aral Sea syndrome desiccates Lake Urmia: Call for action. *Journal of Great Lakes Research, 41*(1), 307–311.

Alborzi, A., Mirchi, A. & Moftakhari, H. et al. (2018) Climate-informed environmental inflows to revive a drying lake facing meteorological and anthropogenic droughts. *Environmental Research Letters* **13** (8).

Azarnivand, A., Hashemi-Madani, F. S., & Banihabib, M. E. (2015). Extended fuzzy analytic hierarchy process approach in water and environmental management (case study: Lake Urmia Basin, Iran). *Environmental Earth Sciences, 73*(1), 13–26.

AZOH. (n.d.) Speech in European Parliament seminar for Urmia Lake: AZOH spokesman. https://azoh.info/index.php?option=com_content&view=article&id=28994:speech-in-european-parliament-seminar-for-urmia-lake-azoh-spokesman&catid=33:english. Accessed 1/27/2024.

Bakhtiari, F. (2020) Lake Urmia in best condition over past decade. https://www.tehrantimes.com/news/446998/Lake-Urmia-in-best-condition-over-past-decade. Accessed 8/10/2024.

Barhagh, S. E., Zarghami, M., Alizade Govarchin Ghale, Y. & Shahbazbegian, M. R. (2021) System dynamics to assess the effectiveness of restoration scenarios for the Urmia Lake: A prey-predator approach for the human-environment uncertain interactions. *Journal of Hydrology* **593**.

Chaudhari, S., Felfelani, F., Shin, S., & Pokhrel, Y. (2018). Climate and anthropogenic contributions to the desiccation of the second largest saline lake in the twentieth century. *Journal of Hydrology, 560*, 342–353.

Danesh-Yazdi, M., & Ataie-Ashtiani, B. (2019). Lake Urmia crisis and restoration plan: Planning without appropriate data and model is gambling. *Journal of Hydrology, 576*, 639–651.

Darling, L. T. (2013). *A history of social justice and political power in the Middle East: The circle of justice from mesopotamia to globalization*. Routledge.

Delju, A. H., Ceylan, A., Piguet, E., & Rebetez, M. (2013). Observed climate variability and change in Urmia Lake Basin, Iran. *Theoretical and Applied Climatology, 111*(1–2), 285–296.

Doorn, N. (2019). *Water ethics: An introduction*. Published by Rowman & Littlefield International Ltd.

Eimanifar, A., & Mohebbi, F. (2007). Urmia Lake (Northwest Iran): A brief review. *Saline Systems, 3*, 5.

FAO & ULRP. (2017). *Integrated programme for sustainable water resources management in Lake Urmia Basin*. https://urmialake.urmia.ac.ir/sites/urmialake.urmia.ac.ir. Accessed 8/18/2025.

Farahmand, H., Tajrishy, M., Isaai, M. T., Ghoreishi, M., & Mohammadi, M. (2023). *How do farmers' perceptions and attitudes toward agricultural water consumption behaviors can lead to unsustainability; evidence from Mahabad plain*. Lake Urmia.

Golabian, H. (2011). Urumia Lake: Hydro-ecological stabilization and permanence. In V. Badescu & R. B. Cathcart (Eds.), *Macro-engineering Seawater in unique environments* (pp. 365–397). Springer.

Grafton, R. Q., Williams, J., & Perry, C. J., et al. (2018). The paradox of irrigation efficiency. *Science (New York, N.Y.)*, **361** (6404), 748–750.

Groenfeldt, D. (2019). *Water ethics: A values approach to solving the water crisis* (2nd ed.). Routledge.

Haghighi, A. T., Fazel, N., Hekmatzadeh, A. A., & Klöve, B. (2018). Analysis of effective environmental flow release strategies for Lake Urmia restoration. *Water Resources Management, 32*(11), 3595–3609.

Hamidi-Razi, H., Mazaheri, M., Carvajalino-Fernández, M., & Vali-Samani, J. (2018). Investigating the restoration of Lake Urmia using a numerical modelling approach. *Journal of Great Lakes Research, 45*(1), 87–97.

Hamshari Online. (2012). Dry Urmia Lake. https://www.hamshahrionline.ir/news/190458. Accessed 08/16/2025.

Hashemi, M. (2012). *A social-technical assessment framework for integrated water resources management (IRWM) in Lake Urmia Basin, Iran* (Doctor of Philosophy Thesis). Newcastle University, Newcastle.

Hassanzadeh, E., Zarghami, M., & Hassanzadeh, Y. (2012). Determining the main factors in declining the Urmia Lake Level by using system dynamics modeling. *Water Resources Management, 26*(1), 129–145.

Hemmati, M., Ahmadi, H., Hamidi, S. A., & Naderkhanloo, V. (2021). Environmental effects of the causeway on water and salinity balance in Lake Urmia. *Regional Studies in Marine Science, 44*.

Hosseini-Moghari, S.-M., Araghinejad, S., Tourian, M. J., Ebrahimi, K., & Döll, P. (2020). Quantifying the impacts of human water use and climate variations on recent drying of Lake Urmia basin: The value of different sets of spaceborne and in situ data for calibrating a global hydrological model. *Hydrology and Earth System Sciences, 24*(4), 1939–1956.

Huang, S., Chen, X., & Chang, C. et al. (2022). Impacts of climate change and evapotranspiration on shrinkage of Aral Sea. *Science of the Total Environment, 845*.

ISNA. (2018). *Restoration of Lake Urmia saves 14 million people—ISNA*. https://en.isna.ir/news/97101005376/Restoration-of-Lake-Urmia-saves-14-million-people. Accessed 1/27/2024.

Karami, N. (2018). The drying of Lake Urmia as a case of the "Aralism" concept in totalitarian systems. *International Journal of Geography and Regional Planning, 4*(1), 43–63.

Karbassi, A., Bidhendi, G. N., Pejman, A., & Bidhendi, M. E. (2010). Environmental impacts of desalination on the ecology of Lake Urmia. *Journal of Great Lakes Research, 36*(3), 419–424.

Kardovani. (n.d.) About Prof. Parviz Kardovani. http://www.kardovani.com/lebenslauf-kardovani-e.htm. Accessed 8/10/2024.

Khalyani, A. H., Mayer, A. L., & Norman, E. S. (2014). Water flows toward power: Socioecological degradation of Lake Urmia, Iran. *Society & Natural Resources, 27*(7), 759–767.

Khatami, S., & Berndtsson, R. (2013). Urmia Lake watershed restoration in Iran: Short- and long-term perspectives *6*.

Latour, B. (2013). *An inquiry into modes of existence: An anthropology of the moderns,* 1st Harvard University Press paperback ed., 1st printing. Harvard University Press, Cambridge (Massachusetts), London (England).

Latour, B. (2017). *Facing gaia: Eight lectures on the new climatic regime; trans. by catherine porter*. Medford, Polity, Cambridge.

Latour, B. (2018). *Existenzweisen: Eine Anthropologie der Modernen* (Erste). Suhrkamp Verlag.

Lüdeke, M. K. b., Petschel-Held, G., & Schellnhuber, H.-J. (2004). Syndromes of global change: The first panoramic view. *GAIA 13*(1).

Madani, K. (2014). Water management in Iran: what is causing the looming crisis? *Journal of Environmental Studies and Sciences 4*(4), 315–328.

Madani, K., AghaKouchak, A., & Mirchi, A. (2016). Iran's socio-economic drought: Challenges of a water-bankrupt nation. *Iranian Studies, 49*(6), 997–1016.

Mardi, A. H., Khaghani, A., MacDonald, A. B., et al. (2018). The Lake Urmia environmental disaster in Iran: A look at aerosol pollution. *The Science of the Total Environment, 633*, 42–49.

Mehr News Agency. (2023). Lake Urmia project proves environment matters much to gov. https://en.mehrnews.com/news/197755/Lake-Urmia-project-proves-environment-matters-much-to-gov. Accessed 8/11/2024.

Micklin, P. (2006). The Aral sea crisis and its future: An assessment in 2006. *Eurasian Geography and Economics, 47*(5), 546–567.

Mirchi, A., Madani, K., & AghaKouchak, A. (2015) Lake Urmia: How Iran's most famous lake is disappearing. https://www.theguardian.com/world/iran-blog/2015/jan/23/iran-lake-urmia-drying-up-new-research-scientists-urge-action. Accessed 10/8/2024.

Moghaddasi, M., Morid, S., Delavar, M., & Hossaini Safa, H. (2017). Lake Urmia Basin drought risk management: A trade-off between environment and agriculture. *Irrigation and Drainage, 66*(3), 439–450.

MoU. (2010). Integrated Management Plan for Lake Urmia Basin. https://www.ramsar.org/sites/default/files/documents/library/lakeurmiamanagementplan-i.r.iran2010.pdf. Accessed 08/16/2025.

Nabavi, E. (2018). Failed policies, falling aquifers: Unpacking groundwater overabstraction in Iran. *Water Alternatives, 11*(3), 699–724.

Nhu, V.-H., Mohammadi, A. & Shahabi, H. et al. (2020) Monitoring and assessment of water level fluctuations of the Lake Urmia and its environmental consequences using Multitemporal Landsat 7 ETM+ Images. *International Journal of Environmental Research and Public Health 17*(12).

Nikraftar, Z., Parizi, E., Hosseini, S. M., & Ataie-Ashtiani, B. (2021). Lake Urmia restoration success story: A natural trend or a planned remedy? *Journal of Great Lakes Research, 47*(4), 955–969.

Opp, C., Abbasi, H., & Banedjschafie, S. (2017). Aral Sea syndrome and Lake Urmia crisis—A comparison of causes, effects and strategies for problem solutions. In A. Dittmann (Ed.) *Geoparks and geo-tourism in Iran*, 34th edn (pp. 169–183). PL Academic Research.

Parsinejad, M., Rosenberg, D. E., & Ghale, Y. A. G. et al. (2022). 40-years of Lake Urmia restoration research: Review, synthesis and next steps. *The Science of the Total Environment, 832.*

Perry, C., Steduto, P., & Karajeh, F. (2017) *Does improved irrigation technology save water?: A review of the evidence.* Discussion paper on irrigation and sustainable water resources management in the Near East and North Africa.

Pouladi, P., Badiezadeh, S., & Pouladi, M. et al. (2021). Interconnected governance and social barriers impeding the restoration process of Lake Urmia. *Journal of Hydrology, 598.*

Presstv. (2023). *President Raeisi inaugurates mega water transfer project to save shrinking Lake Urmia.* https://www.presstv.ir/Detail/2023/02/24/698805/President-Raeisi-inaugurates-mega-water-transfer-project-to-save-shrinking-Lake-Urmia. Accessed 8/11/2024.

Rajabova, S. (2013). *Iran's official says water transfer to Lake Urmia unprofessional.* https://www.azernews.az/region/60438.html. Accessed 8/11/2024.

Ranjbar, A. M. (2022). The greening of human rights in Iran: Lake Orumiyeh, human rights, and environmental justice. *Environment and Planning E: Nature and Space 3*(5).

Rouw M. (2017). *Socially navigating through the Urmia Lake debate: Master Thesis.* https://edepot.wur.nl/421717. Accessed 8/15/2024.

Saemian, P., Elmi, O., Vishwakarma, B. D., Tourian, M. J., & Sneeuw, N. (2020). Analyzing the Lake Urmia restoration progress using ground-based and spaceborne observations. *The Science of the Total Environment, 739.*

Salimi, J., Maknoon, R., & Meijerink, S. (2019). Designing institutions for watershed management: A case study of the Urmia Lake Restoration National Committee. *Water Alternatives, 12*(2), 609–635.

Schmidt, M. (2021). *The Vanishing of Iran's Lake Urmia* (pp. 171–190). Brill, Leiden.

Schulz, S., Darehshouri, S., Hassanzadeh, E., Tajrishy, M., & Schüth, C. (2020). Climate change or irrigated agriculture—What drives the water level decline of Lake Urmia. *Scientific Reports, 10*(1), 236.

Shadkam, S. (2017). *Preserving Urmia Lake in a changing world: Reconciling anthropogenic and climate drivers by hydrological modelling and policy assessment*. Wageningen University.

Shadkam, S., Oel, P., Kabat, P., & Ludwig, F. (2017). *Water-saving interventions assessment framework: an application for the Urmia Lake Restoration Program 19.*

Shadkam, S., van Oel, P., Kabat, P., Roozbahani, A., & Ludwig, F. (2020). The Water-Saving Strategies Assessment (WSSA) framework: An application for the Urmia Lake Restoration Program. *Water, 12*(10).

Shahi, A. (2019). Drought: The Achilles Hell of the Islamic Republic of Iran. *Asian Affairs, 50*(1), 18–39.

Shakoori, A. (2001). *The state and rural development in post-revolutionary Iran.* Palgrave, Houndmills Basingstoke Hampshire.

Shirmohammadi, B., Malekian, A., & Salajegheh, A. et al. (2020) Scenario analysis for integrated water resources management under future land use change in the Urmia Lake region, Iran. *Land Use Policy, 90.*

Sima, S., Rosenberg, D. E., Wurtsbaugh, W. A., Null, S. E., & Kettenring, K. M. (2021). Managing Lake Urmia, Iran for diverse restoration objectives: Moving beyond a uniform target lake level. *Journal of Hydrology: Regional Studies, 35*(9), Article 100812.

Tehran Times. (2018). *Water transfer won't revive Lake Urmia: official*. https://www.tehrantimes.com/news/423519/Water-transfer-won-t-revive-Lake-Urmia-official. Accessed 8/11/2024.

Tehran Times. (2021). *Some $3.5b spent on Lake Urmia revitalization*. https://www.tehrantimes.com/news/463465/Some-3-5b-spent-on-Lake-Urmia-revitalization. Accessed 8/11/2024.

Tehran Times. (2023). *Special budget to expedite Lake Urmia restoration*. https://www.tehrantimes.com/news/480997/Special-budget-to-expedite-Lake-Urmia-restoration. Accessed 8/11/2024.

Transiskus, S. (2021). Umweltmigration und Immobilität am Urmiasee (Iran). In M. Schmidt & H. Zapf (Eds.), *Environmental humanities: Beiträge zur geistes- und sozialwissenschaftlichen Umweltforschung* (pp. 218–316).

Tourian, M. J., Elmi, O., Chen, Q., Devaraju, B., Roohi, S., & Sneeuw, N. (2015). A spaceborne multisensor approach to monitor the desiccation of Lake Urmia in Iran. *Remote Sensing of Environment, 156*, 349–360.

Tussupova, K., Anchita, A., Hjorth, P., & Moravej, M. (2020). Drying Lakes: A Review on the Applied Restoration Strategies and Health Conditions in Contiguous Areas. *Water, 12*(3), 749.

ULRP. (2015). *Urmia Lake restoration program: Brief report and projects outline*.

ULRP. (2018). *Urmia Lake: lessons and challenges*.

United Nations Environmental Programme (UNEP) & Global Environmental Alert Service (GEAS). (2012). The drying of Iran's Lake Urmia and its environmental consequences. *Environmental Development, 2*, 128–137.

UNPO. (2015). *Conference at European Parliament Addresses Lake Urmia's Alarming Deterioration*. https://unpo.org/conference-at-european-parliament-addresses-lake-urmia%ef%bf%bds-alarming-deterioration/. Accessed 1/27/2024.

Vaheddoost, B., & Aksoy, H. (2018). Interaction of groundwater with Lake Urmia in Iran. *Hydrological Processes 9*(2).

Su, Y., Li, X., & Feng, M. et al. (2021). High agricultural water consumption led to the continued shrinkage of the Aral Sea during 1992–2015. *Science of the Total Environment, 777*.

Zarghami, M., & AmirRahmani, M. (2017). A system dynamics approach to simulate the restoration plans for Urmia Lake, Iran. In A. Matsumoto (Ed.), *Optimization and dynamics with their applications* (pp. 309–326). Springer Singapore.

Ženko, M. (2020). *Hydro-social transformations in the Lake Urmia basin, Iran*.

Ženko, M., & Menga, F. (2019). Linking water scarcity to mental health: Hydro–social interruptions in the Lake Urmia Basin, Iran. *Water, 11*(5).

Zoljoodi, M., & Didevarasl, A. (2014). Water-level fluctuations of Urmia Lake: Relationship with the long-term changes of meteorological variables (solutions for water-crisis management in Urmia Lake Basin). *Atmospheric and Climate Sciences, 04*(03), 358–368.

CHAPTER 8

Beyond Water: The Multiple Natures and Contested Futures of Lake Urmia

What is Lake Urmia? Sure, it is a lake, a basin full of water, and at first, the question seems like an unnecessary tautology. Nevertheless, as I hope this thesis has established, there is much more to a lake than 'just' water and its materiality. What is more, H_2O may not be even the only central actor but one of many others in this case. In order to understand, what constitutes Lake Urmia, it was suggested by the hydro-social cycle to extend the circle of entities by including social, political and technological dimensions that have been relevant in the lake's mutation. In addition, the cartography of controversy provided a way to deepen ontological inquiry, as recommended by Linton's (2010) hydrolectics. Yet, perhaps even more importantly, one of the pillars of this research was the emphasis on process; that is to say, the historical, political and—in this case primarily—the scientific processes and discourse that have shaped Lake Urmia and people's understanding thereof.

Water in the face of Iran's history has always been a central actor, whose social nature has become self-evident. Especially in the Iran of the Achaemenid, Sassanid and the first Arab caliphates, water was not only the central pillar of the economy, but it was also an anthropomorphised, subjectivised entity. The introduction of Islam, in combination with its increasing emphasis on human needs and the hydraulic mission in the twentieth century with its techno-managerial paradigm, caused a radical ontological shift. Water and associated infrastructure such as

R. Gonda, *Lake Urmia's Hydro-Social Dilemma*,
https://doi.org/10.1007/978-3-032-08941-0_8

large-scale dams, water transfer and diesel-pumped wells and centralised and Westernised water bureaus were antithetical to the traditional water management system. The *buneh* system, qanats and the emphasis on agriculture as the economic foundation of Iran are nowadays relics, almost anachronisms, that could be found at Lake Urmia but are increasingly pushed back by modern water and the political as well as technological transformation that follows. Iran has moved its interest to oil exports and the industrial sector. Agriculture has turned from the country's social and economic backbone to a floating state, whereby it is stuck between the failed hydraulic mission, especially in the aftermath of the Islamic Revolution, and the need to *ecologise* (instead of modernise) in order to be in a sustainable state with other non-human actants. To be more specific, given the example of irrigation efficiency above, Iran's water management has failed to include other non-human entities such as water for soil or to recharge the water table, even when it is driven by a sustainable paradigm; this, however, is only superficially the case. The historical process, as explained in detail, has taken part in the mutation of Lake Urmia, clarifying that the metamorphosis has been the result of numerous actions that even extend into today's Iran.

Inquiries into the scientific landscape at Lake Urmia confirmed the lopsided composition of knowledge production. The movement of water and its management are prioritised, leaving a gap in political and social, human factors of the crisis. How does it come that during the—not yet—declared Anthropocene, even at Lake Urmia the Anthropos has almost left the scene and turned into a lifeless backdrop, while even the brine shrimp and its needs have been given more attention? It is almost paradoxical that in this case nature has been vivified and, generally speaking, social factors muted and objectified. However, the sciences acknowledge this 'data' gap, and an increasing number of scientists have turned their attention to social factors and the investigation of, for example, political institutions and farmers' struggles on the ground, even though this always bares the risk of politicising Lake Urmia, which is a difficult task in Iran, as it forces especially local scientists to tiptoe around clear statements. In contrast, (Iranian) scientists employed at foreign universities have attempted to blame the disaster on political misguidance and myopic water management. This criticism has mostly been formulated by Iranian scientists at Anglo-American universities, with Madani at the forefront, followed by others such as AghaKouchak and Pouladi. They insist on driving, or better, relocating the prima causa of Lake Urmia's mutation at the centre of Iran, its government and the lack of sustainable praxis on the ground. This position aims to prohibit any attempt, as has been the case especially during Ahmadinejad's presidency, to move the responsibility primarily to exogenous factors such as climate change (a global responsibility) and geopolitical factors (a foreign country or countries' responsibility). This push towards the anthropogenic drought has also been shared by the synthesis report of Parsinejad et al. (2022), which was also identified as a panorama—to make use of the ANT jargon. However, this panorama revealed some issues in its presentation of this debate, as the climatic or exogenous factors were rather belittled by offering questionable numbers and ignoring qualitative, non-numerical statements (e.g., Schulz et al., 2020). The debate on the main cause of Lake Urmia's desiccation was in my estimation prematurely short-circuited and proposed as first-degree objectivity—this is not to say that it will not reach this state in the near future but it should be left as second-degree objectivity for now. Nevertheless, the more the scientific discourse moves to identify the 'nature' of the crisis as anthropogenic (whereby most scientists declare agriculture as the main issue), the more demand-focused solutions are perceived as obligatory, because the closer the main cause is located to the lake and its basin, the more farmers and institutions are held accountable for reducing water consumption sustainably. Therefore, it is not a coincidence that most scientists recommend demand-focused solutions, while especially supply-focused and alternative solutions are less mentioned.

The analysis of the four controversies revealed a diverse set of water concepts. First, the controversies of the minimal ecological water level and the causeway prove the point made above this conclusion: according to most scientists, salt and *A. urmiana* should be considered central actors in the hydro-social dynamics at Lake Urmia, and therefore they co-constitute the lake. The reason for this thinking lies in the far-reaching consequences that come with these two actors. They argue that if salt concentrations in Lake Urmia exceed a certain limit, then the brine shrimp will likely go extinct, which in consequence will lead to significant environmental degradation. As a result, the majority of scientists and the ULRP have both demanded and aligned their goals according to the minimal ecological water level at 1,274.1 m, which should guarantee to keep salt, or more specifically NaCl concentrations, below 240 g/l. Thus, the *A. urmiana*, salt and water (volume) triangle has played a central role in the restoration efforts primarily made by the ULRP. The priority of the minimal ecological water level emphasises a type of water that allows *A.*

urmiana to reproduce sustainably, since it is the basis of the flat hierarchy of the lake's aquatic ecology. Water is meant to achieve life-sustaining status, i.e., an environment for the brine shrimp, which would extend to an improvement in environmental health beyond the lake (e.g., marshes and islands). If salt concentrations exceed their limits, the lake will turn into a dead, lifeless body, as well as the adjacent environment. A sufficient amount of water is meant to 'do it all'. Conversely, some criticism from scientists has emerged in the last couple of years. According to them, the goal to reach the minimal ecological water level is too narrow-minded and might not bring the promised restoration of the lake as a holistic ecological, economic and social entity. They demand multiple indicators and a flexible water volume supply according to the given anthropogenic and climatic conditions in the basin. Otherwise, the planned restoration is likely to fail in face of circumstantial dynamics. Moreover, the legitimacy of the minimal ecological water level is in question, as the 240 g/l-limit has not been proved empirically but is based on anecdotal evidence according to its opponents. This goal discrepancy was also registered in the Memorandum of Understanding (2010) and the ULRP (2015; 2018). While the former focused entirely on the ecological water level and aligned all of its projects in accordance with a fixed water volume, the latter approached its restoration with a 25-year vision—a fictional, ideal Lake Urmia landscape (where the water level was one of many targets). To put it simply, water is either omnipotent for some actors or it is just one of many factors necessary to ensure Lake Urmia's restoration or revitalisation for others. What they all seem to agree on, though, is that water serves as a conduit to improve the lake's ecological conditions.

The causeway controversy, on the other hand, has disclosed the many faces of this infrastructure. The majority of scientists, including the media and public opinion, perceive and declare the causeway—part bridge, part dam—as a contributor to the lake's degradation, especially its ecology, since they assume that salt concentrations rise due to a lack of water circulation. For some, the causeway might be even the main reason for this mutation. In this context, two water concepts were identified that carry shades of modern water but should be differentiated in certain aspects. First, there is the water concept of the scientists, which promotes a sensitive relation between the lake and the causeway, which is similar to the public's local-traditional, and in many ways religious, water concept. The causeway has hurt Lake Urmia's natural unity, divided it, disrupted its functionality and destroyed its architecture. The public describes how the causeway has destroyed underground water caverns that should re-fill the lake. The scientists may disagree with this not empirically proven aspect of the public's concept; nevertheless, there is a high probability that they established their water concept on the *pristine nature* concept of the public. This means that any human intervention in the natural (water) cycle is by default suspected of causing harm. My assumption could be substantiated by the fact that numerous papers misused certain references to shed a negative light on the causeway, although evidence up to 2014 suggested no significant negative effect on the lake's ecology. To diversify perspectives on the causeway, other scientists have even suggested a positive relation between it and Lake Urmia—something that can be understood as 'symbiotic', not in a biological sense but in a social one. As the causeway might reduce salt concentrations in favour of the brine shrimp, it co-produces a healthy environment and as such turns from an antagonist in the debate to a supporting cast member. In this sense, it would be a welcome actor and part of the hydro-social collective of Lake Urmia and its basin. Only a minority of scientists have taken a neutral position on the causeway, whereas the empirically based discussion is yet to find momentum, since it needs more research on this still controversial matter (see overview of controversies Fig. 8.1).

Lake Urmia is not just a lake. It is also an idea. Moreover, how current and future generations will talk about and debate will have an impact on how collective life will take place in the basin. The sciences have a responsibility in this regard, as much as the economists, politicians, farmers, engineers and many more. Although Lake Urmia has lost its title as the region's turquoise solitaire, it will survive as an ideal in people's minds, and long as this is the case, there is always a chance of a new lake appearing that indicates an adaptive human and non-human society.

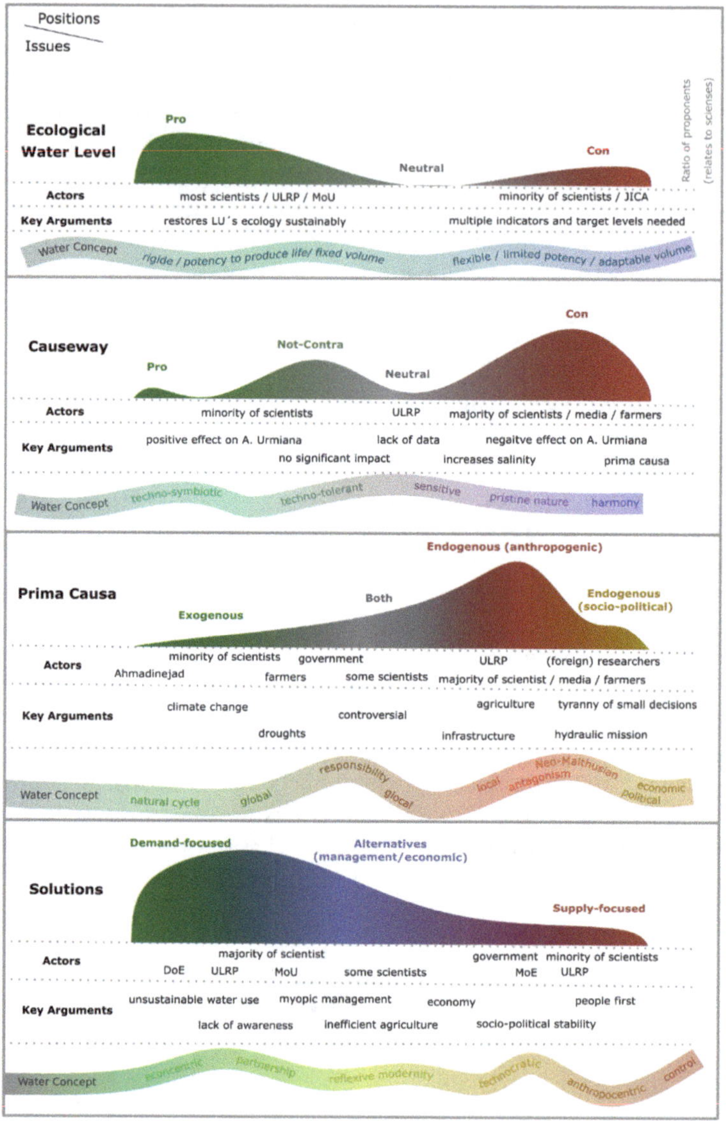

Fig. 8.1 Overview of controversies—positions, actors, water concepts, key arguments

References

Linton, J. (2010). *What is water?: The history of a modern abstraction*. UBC Press.

MoU. (2010). *Integrated Management Plan for Lake Urmia Basin*. https://www.ramsar.org/sites/default/files/documents/library/lakeurmiamanagementplan-i.r.iran2010.pdf. Accessed 08/16/2025.

Parsinejad, M., Rosenberg, D. E., Ghale, Y. A. G., et al. (2022). 40-years of Lake Urmia restoration research: Review, synthesis and next steps. *The Science of the Total Environment, 832*, Article 155055.

Schulz, S., Darehshouri, S., Hassanzadeh, E., Tajrishy, M., & Schüth, C. (2020). Climate change or irrigated agriculture—What drives the water level decline of Lake Urmia. *Scientific Reports, 10*(1), 236.

ULRP. (2015). *Urmia Lake restoration program: Brief report and projects outline*.

ULRP. (2018). *Urmia Lake: Lessons and challenges*.

Index

R. Gonda, *Lake Urmia's Hydro-Social Dilemma*, https://doi.org/10.1007/978-3-032-08941-0

FSC
www.fsc.org
MIX
Papier | Fördert gute Waldnutzung
FSC® C083411

Kolibri 360 GmbH
Ferdinand-Jühlke-Straße 7
99095 Erfurt, Deutschland
produktsicherheit@kolibri360.de